SUFFICIENT STATISTICS

STATISTICS

Textbooks and Monographs

A SERIES EDITED BY

D. B. OWEN, Coordinating Editor
Department of Statistics
Southern Methodist University
Dallas, Texas

PAUL D. MINTON
Virginia Commonwealth University
Richmond, Virginia

JOHN W. PRATT
Harvard University
Boston, Massachusetts

Volume 1: The Generalized Jackknife Statistic, *H. L. Gray and W. R. Schucany*

Volume 2: Multivariate Analysis, *Anant M. Kshirsagar*

Volume 3: Statistics and Society, *Walter T. Federer*

Volume 4: Multivariate Analysis: A Selected and Abstracted Bibliography, 1957-1972, *Kocherlakota Subrahmaniam and Kathleen Subrahmaniam*

Volume 5: Design of Experiments: A Realistic Approach, *Virgil L. Anderson and Robert A. McLean*

Volume 6: Statistical and Mathematical Aspects of Pollution Problems, *John W. Pratt*

Volume 7: Introduction to Probability and Statistics (in two parts) Part I: Probability; Part II: Statistics, *Narayan C. Giri*

Volume 8: Statistical Theory of the Analysis of Experimental Designs, *J. Ogawa*

Volume 9: Statistical Techniques in Simulation (in two parts), *Jack P. C. Kleijnen*

Volume 10: Data Quality Control and Editing, *Joseph I. Naus*

Volume 11: Cost of Living Index Numbers: Practice, Precision, and Theory, *Kali S. Banerjee*

Volume 12: Weighing Designs: For Chemistry, Medicine, Economics, Operations Research, Statistics, *Kali S. Banerjee*

Volume 13: The Search for Oil: Some Statistical Methods and Techniques, *edited by D. B. Owen*

Volume 14: Sample Size Choice: Charts for Experiments with Linear Models, *Robert E. Odeh and Martin Fox*

Volume 15: Statistical Methods for Engineers and Scientists, *Robert M. Bethea, Benjamin S. Duran, and Thomas L. Boullion*

Volume 16: Statistical Quality Control Methods, *Irving W. Burr*

Volume 17: On the History of Statistics and Probability, *edited by D. B. Owen*

Volume 18: Econometrics, *Peter Schmidt*

Volume 19: Sufficient Statistics: Selected Contributions, *Vasant S. Huzurbazar (edited by Anant M. Kshirsagar)*

OTHER VOLUMES IN PREPARATION

SUFFICIENT STATISTICS

SELECTED CONTRIBUTIONS

Vasant S. Huzurbazar
Department of Mathematics and Statistics
University of Poona
Poona, India

edited by Anant M. Kshirsagar
Institute of Statistics
Texas A&M University
College Station, Texas

MARCEL DEKKER, INC. New York and Basel

MARCEL DEKKER, INC.
270 Madison Avenue, New York, New York 10016

LIBRARY OF CONGRESS CATALOG CARD NUMBER: 76-1575
ISBN: 0-8247-6296-7

Current printing (last digit):
10 9 8 7 6 5 4 3 2 1

PRINTED IN THE UNITED STATES OF AMERICA

To My Wife

PRABHA

FOREWORD

I have great pleasure in introducing this book by Professor V. S. Huzurbazar to the readers. Professor Huzurbazar's fundamental contributions in the areas of sufficiency, invariance, and maximum likelihood estimation during the last 20 years or so are well-known to the statisticians. Part I of this book brings together his research contributions on Bayesian inference and invariance theory while Parts II and III deal with sufficient statistics.

The pendulum of opinion about Bayesian inference which was once on the rejection side is now between acceptance and rejection. Statisticians have now realized that Bayesian inference is not as unsatisfactory as they once thought and the alternatives to it are not as satisfactory as they were proclaimed to be. This change in the status of Bayesian techniques is due to a large number of devoted workers in this area, Sir Harold Jeffreys being one of the leaders among them. Professor Huzurbazar was fortunate to be able to work with him at Cambridge. It is no wonder that Professor Huzurbazar was able to contribute substantially in this field. His work, however, was not easily accessible to students and teachers of statistics. Professor Huzurbazar is known for his love of rigorousness and was reluctant to sacrifice rigor in order to cut down the length of his research papers. Scientific journals, unable to publish long articles, could not accommodate his contributions completely. It was a pity, therefore, that such valuable work, which was awarded the coveted Adam's prize by Cambridge University (England) remained comparatively inaccessible. Professor Huzurbazar was busy all these years guiding his own research students and building up a fine school of statistics at Poona University (India) and did not find any time to remedy this situation. On one of my visits to India, Dr. Maurits Dekker of Marcel Dekker, Inc. requested that I be his ambassador-at-large for finding good potential authors in statistics in India and naturally the first name that occurred to me was that of Professor Huzurbazar. He kindly agreed to hand over his collection of material to me so that it could be brought out as a monograph.

There is a wealth of information in this collection. Students and teachers alike will find this monograph very interesting and useful. While teaching courses in statistical inference, I found that the treatment given to the "nonregular case" insufficiency in almost all the standard textbooks was very unsatisfactory. I did not find any completely satisfactory proof of the theorem that the general form of distribution in the nonregular case admitting a sufficient statistic is $g(x)/h(\theta)$. Many points were either ignored or left to intuition. The first time I found a completely rigorous treatment of this was when I read Dr. Huzurbazar's manuscript. The same is also true about the joint distribution of the smallest, the greatest, and any unordered member in samples from a distribution. Most texts simply present the final formula but Professor Huzurbazar spells out all the details very rigorously.

Another interesting feature of this monograph is the variety of illustrative examples. In addition to the routine examples of Normal, Cauchy, exponential distributions, Dr. Huzurbazar gives a large number of other distributions also. This will be especially useful to teachers who on occasion are likely to be bored with the same type of illustrations in all the texts.

Part I of this book, which gives a detailed account of Jeffreys' invariance theory of prior possibilities, will go a long way in meeting some of the objections raised by the opponents of Bayesian inference. Professor Huzurbazar's own contributions about the properties of Jeffreys' invariants are also valuable and the readers will get a full account of all this in a unified and collected form for the first time.

Professor Huzurbazar is a gifted teacher and explains difficult concepts with ease. His interesting analogy of the prior probabilities with the initial starts given to athletes in a running race is an example of this.

I am sure this monograph will be very useful to students and research workers in statistics. It is now commonly admitted that Bayesian analysis, by a careful choice of prior distributions, can throw new light on the data and can be helpful. It is therefore necessary to make an attempt to bring all the relevant aspects of this area and other related works together and consider its wider aspects. Professor Huzurbazar's monograph will go a long way in this direction.

A. M. Kshirsagar
Texas A & M University
College Station, Texas

PREFACE

This book is a collection of my research monographs dealing mainly with properties of sufficient statistics.

Part 1 of this book is my research work on Bayesian inference and invariance theory of prior probabilities, and it was awarded the Adam's Prize for the years 1959 and 1960 by the University of Cambridge, England. I am grateful to Professor Sir Harold Jeffreys, F.R.S. of Cambridge University for initiating me into the subject of inverse probability when I was a research student at Cambridge (1946-49).

Parts 2 and 3 of this book are two research monographs on "The general forms of distributions admitting sufficient statistics for parameters in nonregular cases" and "Location and scale parameters and sufficient statistics." This research work was sponsored by the National Science Foundation of the United States when I was a Fulbright Visiting Professor in the Department of Statistics, Iowa State University of Science and Technology, Ames, Iowa (1962-64). I am thankful to these organizations for their support of my work and I am also thankful to Professor T. A. Bancroft, the then Director and Head of the Department of Statistics at Iowa State University.

Editors of some current journals in statistics appreciated the importance of results obtained in these monographs. However, on account of limitations of space in their journals, they advised me to publish the monographs in a book form. Although unpublished, some results in these monographs have been mentioned by some authors in their published papers and books. In particular, some results in the invariance theory of prior probabilities have been extensively referred to by Professor Jeffreys in the Third Edition (1961) of his book "Theory of Probability" (Oxford University Press). Some remarks made by me on establishment of sufficiency of a statistic in Part I of the book, have been named as "Huzurbazar's Conjecture" by Dr. C. R. Rao, Dr. Yu V. Linnik and Dr. V. N. Sudakov, as mentioned in the Second Edition of "Linear Statistical Inference and Its Applications" by C. R. Rao (John Wiley and Sons, 1973).

I wish to express my deepest gratitude to Professor A. M. Kshirsagar of the Institute of Statistics, Texas A&M University, for taking a lead in the publication of this book, for sparing his valuable time in supervising the work of typing the manuscript of the book and for his valuable advice. I am also thankful to Professor D. B. Owen for his encouragement and useful advise in the publication of this book. Finally I wish to thank Marcel Dekker, Inc. for their keen interest in my work, and Mrs. Karen Holder, who did the typing with great speed and accuracy.

V. S. Huzurbazar
Department of Mathematics and Statistics
University of Poona
India

CONTENTS

SUFFICIENT STATISTICS

ERRATA

p. 50, 3rd line from bottom: Read "Type II^1" in place of "Type III"
p. 53, 3rd line from top: Read "standard" in place of "second"
p. 85, Example 2, last line: Read "$d\alpha/\alpha$, α" in place of "$d\alpha/\alpha$,"
p. 86, line following Eq. (7-16): Read "invariants stated" in place of "invariance states"
p. 89, Sec. 8.4.3, 2nd line: Read "$\frac{e^{-\alpha}\alpha^{x}}{x!}$" in place of "$\frac{e^{-\alpha x}}{x!}$"
p. 96, Contents, Sec. 3.4, 2nd line: Read "Eq. (3-26)" in place of "$\Theta \geq T \geq b_1^{-1}[x_{(n)}]$"
p. 96, Contents, Sec. 3.5, 3rd line: Read "Eq. (3-26)" in place of "$\Theta \geq T \geq b_1^{-1}[x_{(n)}]$"
p. 96, Contents, Sec. 3.6, 4th line: Delete "$\Theta \geq T \geq b_1^{-1}[x_{(n)}]$"
p. 99, 1st line from top: Read "Theorem 7 of Sec. 3.10" in place of "Theorem 2 of Sec. 3.9"
p. 99, 12th line from top: Read "Sec. 2.4" in place of "Sec. 2.3" and read "Sec. 2.6" in place of "Sec. 2.5"
p. 99, 3rd line from bottom: Read "Sec. 2.10" in place of "Sec. 2.9"
p. 101, 6th line from top: Read "Sec. 6" in place of "Sec. 5"
p. 102, 2nd line from bottom: Read "Theorem 6" in place of "Theorem 2"
p. 117, 10th line from top: Read "$P[x_1 = x_2 = \dots = x_i = \dots = x_n]$" in place of "$[x_1 = x_2 = \dots = x_i = \dots = x_n]$"
p. 128, 8th line from top: Read "Eq. (3-3)" at end of line
p. 133, Sec. 3.4, 2nd line: Delete "$\Theta \geq T \geq b_1^{-1}[x_{(n)}]$,"
p. 142, Sec. 3.5, 2nd line: Delete ", $\Theta \geq T \geq b_1^{-1}[x_{(n)}]$"
p. 143, 4th line from top: Read "R_{T,x_i}" in place of "B_{T,x_i}"
p. 147, Sec. 3.6, 2nd line: Delete ", $\Theta \geq T \geq b_1^{-1}[x_{(n)}]$"
p. 149, line following Eq. (3-111): Read "ρ nondegenerate" in place of "nondegenerate"
p. 156, Eq. (3-134): Read "$[a_1^{-1}[\phi_1(T)], a_2^{-1}[\phi_1(T)]]$" in place of "$a_1^{-1}[\phi_1(T)], a_2^{-1}[\phi_1(T)]$"
p. 156, Eq. (3-136): Read "$[b_1^{-1}[\phi_2(T)], b_2^{-1}[\phi_2(T)]]$" in place of "$b_1^{-1}[\phi_2(T)], b_2^{-1}[\phi_2(T)]$"
p. 157, line following Eq. (3-141): Read "Sec. 2.4" in place of "Sec. 2.3"
p. 198, Eq. (2-5): Read "$x > b$" in place of "$x < b$"
p. 199, 3rd line from bottom: Read "$f(x; \theta)$" in place of "$(x; \theta)$"
p. 205, Eq. (2-23): Insert left parenthesis on the right-hand side of the equation to read "$(k_1$" in place of "k_1"
p. 219, 2nd line from bottom: Read "(5-10)" in place of "(5-12)"
p. 250, Eq. (10-16): Delete "σ" from the denominator on the right-hand side of the equation
p. 250, Eq. (10-17): Read "$b_0^{c_2+1}\sigma$" in place of "$b_0^{c_2+1}$" on the right-hand side of the equation
p. 258, Eq. (11-24): Put parentheses on the right-hand side of the equation to read "$\exp(c_0x^2 + c_1x + c_2)$" in place of "$\exp c_0x^2 + c_1x + c_2$"
p. 260, Sec. 11.3.2, 5th line: In the lower limit of the second integral, read "0" in place of "00"
p. 261, References, 3rd line from bottom: Read "Technical report submitted to the National Science Foundation of the U.S.A." in place of "J. R. Stat. Soc. B"

PART 1

INVARIANCE THEORY OF PRIOR PROBABILITIES

1. Bayes' Theorem and Inverse Probability

1.1 Introduction

The main object of Part 1 of this book is to contribute to the invariance theory of prior probabilities originated by Jeffreys (1946, 1948). It is beyond the scope of this essay to discuss various theories of probability and their criticisms. For our purposes we shall assume the axiomatic approach to probability as developed by Jeffreys (1948). The fundamental idea in his theory is that of a reasonable degree of belief intermediate between proof and disproof. It is an extension of ordinary logic which deals with the extreme cases only. Jeffreys assumes that for a given proposition q and for a given data p, there is only one reasonable degree of belief. He emphasizes that the fundamental idea is not simply the probability of a proposition q, but the probability of q on data p. The probability is thus a function of two arguments, both propositions. We shall use the following notation of Jeffreys.

$P(q|p)$ denotes the probability of a proposition q on data p. H stands for previous information. θ will be used to represent the observational data. For a continuous variate x, $P(dx|p)$ is the probability that x lies in a particular range $(x - \frac{1}{2}\,dx,\ x + \frac{1}{2}\,dx)$ of length dx, given the data p. If $x_1, x_2, \ldots, x_n$ are a set of variables, for brevity we shall write (x_j) for $(x_1, x_2, \ldots, x_n)$ as an argument of a function. Thus $\phi(x_1, x_2, \ldots, x_n) = \phi(x_j)$.

1.2 Bayes' Theorem

Fundamental to the theory of inverse probability is the celebrated theorem of Bayes (1763). If $q_1, q_2, \ldots,$ are a set of alternatives, H the previous information, and θ some additional information, then the rule of multiplication of probabilities gives

$$P(\theta q_r|H) = P(\theta|H)\ P(q_r|\theta H) = P(q_r|H)\ P(\theta|q_r H) \tag{1-1}$$

whence

$$P(q_r|\theta H) = \frac{P(q_r|H)\ P(\theta|q_r H)}{P(\theta|H)}. \tag{1-2}$$

In Eq. (1-2), the q_r need not be mutually exclusive or exhaustive. If the q_r are a mutually exclusive and exhaustive set,

$$P(\theta|H) = \sum_r P(\theta q_r|H)$$

$$= \sum_r P(q_r|H)\ P(\theta|q_r H). \tag{1-3}$$

Then Eq. (1-2) becomes

$$P(q_r|\theta H) = \frac{P(q_r|H)\ P(\theta|q_r H)}{\sum_r P(q_r|H)\ P(\theta|q_r H)}, \tag{1-4}$$

which is Bayes' Theorem (or Bayes' Formula, or Bayes' Rule, as it is sometimes called) in its familiar form. It is an immediate consequence of the product rule.

If the q_r are regarded as a set of hypotheses and θ represents a set of observations, $P(q_r|H)$ is called the "prior probability" of q_r, while $P(q_r|\theta H)$ is the "posterior probability" of q_r. $P(\theta|q_r H)$ is the "likelihood" of q_r, a convenient term introduced by Fisher (1922). It is the probability of the observations, given the original information and the hypothesis under consideration. The term "a priori probability" is sometimes used for the prior probability, but as Jeffreys remarks, this term has been used in so many senses that the only solution is to abandon it.

Formula (1-4) then gives the posterior probabilities of the hypotheses q_r in terms of their prior probabilities and likelihoods. Since the denominator of the fraction on the right-hand side of Eq. (1-2) or (1-4) is the same for all q_r, we can write Eq. (1-2) or (1-4) for variations of q_r in the convenient form given by Jeffreys:

$$P(q_r|\theta H) \propto P(q_r|H)\ P(\theta|q_r H). \tag{1-5}$$

This is the "principle of inverse probability." In words: "The posterior probabilities of the hypotheses are proportional to the products of the prior probabilities and the likelihoods." It is the chief rule involved in the process of learning from experience. Jeffreys (1957, p. 30) remarks: "This theorem (due to Bayes) is to the theory of probability what Pythagoras' theorem is to geometry." If the q_r are a mutually exclusive and exhaustive set, as in Eq. (1-4), the constant of proportionality in Eq. (1-5) may also be determined by the condition that the sum of posterior probabilities of all q_r is unity. The principle also covers the cases when the number of hypotheses is infinite. But as remarked by Jeffreys, it may be more convenient in some of these cases to represent certainty by infinity, rather than by the conventional number unity.

1.3 Estimation of Parameters by Inverse Probability

Let $f(x, \alpha)$ be the p.d.f. (probability density function) of a variate x depending on an unknown parameter α. If x is discrete, $f(x, \alpha)$ is the probability function of x. Let $P(d\alpha|H) = \phi(\alpha)\, d\alpha$ be the probability differential of the prior-probability distribution of α. This is the prior probability of the hypothesis that the unknown value of α lies in the range $(\alpha - \frac{1}{2}\, d\alpha, \alpha + \frac{1}{2}\, d\alpha)$ of length $d\alpha$. If α is discrete, we have to write $P(\alpha|H) = \phi(\alpha)$ as the prior probability of the value α. θ denotes the observational data, which consists here of a random sample $x_1, x_2, \ldots, x_n$ of n independent observations from the distribution. Then $P(d\alpha|\theta H)$ is the probability differential of the posterior probability distribution of α, given the sample. The likelihood of the hypothesis corresponding to $d\alpha$ (or α when it is discrete) is $\prod_{i=1}^{n} f(x_i, \alpha)$.

By the principle of inverse probability,

$$P(d\alpha|\theta H) \propto \phi(\alpha)\, d\alpha \prod_{i=1}^{n} f(x_i, \alpha),$$

i.e.,

$$P(d\alpha|\theta H) = k\, \phi(\alpha) \prod_{i=1}^{n} f(x_i, \alpha)\, d\alpha. \tag{1-6}$$

Then $\int_{(\alpha)} P(d\alpha|\theta H) = 1$ gives

$$k = \frac{1}{\int_{(\alpha)} \phi(\alpha) \prod_{i=1}^{n} f(x_i, \alpha)\, d\alpha}, \tag{1-7}$$

the integral being taken over the range of the possible values of α. k will be a function of $x_1, x_2, \ldots, x_n$. Thus Eq. (1-6) gives the complete posterior-probability distribution of the unknown parameter α, given the observations θ.

The method can be easily extended to the case when there are several unknown parameters or when there are several variates.

1.4 Criticism of Bayes' Theorem

We shall not delve deeply into the subtleties of prolonged controversies over the Bayes' Theorem. The practical use of Bayes' Theorem depends on a knowledge of the prior probabilities of hypotheses. When the prior probabilities are not known, and where we have no ground whatever for assuming them to be different, Bayes proposed that the prior probabilities should be taken to be equal. This is called in the literature by various names such as "Bayes' dictum," "Bayes' postulate," the "principle of insufficient reason" (Laplace), or the "equal distribution of ignorance." Though Bayes is careful in his memoir to stress that the principle should be used only in cases where we have no ground whatever for choosing between the hypotheses, his successors (including Laplace) have applied it indiscriminately, assuming equal distribution of the prior probability without justification. It is known that indiscriminate use of this principle leads to unsatisfactory results or contradictions. Fisher (1944, p. 9) remarks that the theory of inverse probability is founded upon an error, and must be wholly rejected. But he seems to recognize the use of inverse probability as legitimate when the prior probabilities are known, though he refers to such cases as trivial.

It is now generally agreed that Bayes' Theorem, and other consequences derived from it, are necessary consequences of fundamental concepts and theorems of the theory of probability. Once we admit these fundamentals, we must admit Bayes' Theorem and all that follows from it. In terms of known prior probabilities, the theorem is indisputable in any theory of probability. It is over the postulate to be adopted when the prior probabilities are not known that all the difficulties and controversies arise. Even Neyman (1942), who is one of the leading exponents of direct systems devised to sidestep use of inverse probability, accepts Bayes' Theorem as legitimate when the prior probabilities are known.

Jeffreys, who is the principal modern exponent of inverse probability, has done much to rehabilitate the theory of inverse probability and show its simplicity and power in dealing with modern statistical problems. For any assessment of the prior probability, the principle of inverse probability will give a unique posterior probability. This can be used as the prior probability in taking account of a further set of data, and the theory can therefore always take account of new information. The choice of the prior probability at the outset, that is, before taking into account any observational information at all, requires of course some consideration.

It is interesting to consider a few special cases which bring out the relationship between the principle of inverse probability and scientific inference.

If all the hypotheses have the same prior probability, their posterior probabilities will be proportional to their likelihoods. The hypothesis with maximum likelihood will then have the maximum posterior probability. In this case, Fisher's celebrated principle of maximum likelihood for choosing among a set of hypotheses is equivalent to the principle of maximum posterior probability.

If all the hypotheses have the same likelihood, it follows from Eq. (1-4) that their posterior probabilities are the same as their prior probabilities. The hypothesis with maximum prior probability will then have the maximum posterior probability. In other words, the observations tell us no new thing. This is the case of irrelevance (Jeffreys, 1957, p. 30).

If $P(q_r|H) = 0$, then $P(q_r|\theta H) = 0$ also. Hence, if a hypothesis is already known to be false, it will still be known to be false, given any additional evidence.

It follows that the posterior probability of a hypothesis can hardly ever be unity, since this would require that all the terms except one in the denominator of Eq. (1-4) be zero. But if one of the hypotheses, say q_1, has a large likelihood and the likelihoods of the remaining hypotheses are small, and the prior probabilities of all the hypotheses are comparable, then the posterior probability of q_1 may be near unity. This is the type of inference known as a crucial test (Jeffreys, 1957, p. 30).

Suppose that the likelihood of q_1 is unity, and that nevertheless $\sim\theta$ (negation of θ) is verified. Then $P(q_1|\sim\theta \cdot H) = 0$. As pointed out by Jeffreys, this explains how failure of a crucial test may lead a previously plausible hypothesis to be rejected. The principle of inverse probability therefore accounts for the use of the most striking type of experiment, the crucial test.

Thus the principle of inverse probability leads to sensible results in the special cases mentioned above. There are many other applications, and a general theory of induction is impossible without it.

An objection raised by the opponents of inverse probability is that when the prior probabilities are not known, they are chosen arbitrarily and introduce an arbitrariness into the whole theory. When the prior probabilities of the hypotheses are not exactly known at the outset, it is possible in some situations to know them approximately. It is also possible to obtain fairly satisfactory mathematical forms expressing the fact that the prior probabilities are unknown. As Jeffreys (1948, p. 102) emphasizes, "... a probability is merely a number associated with a degree of reasonable confidence and has no purpose except to give it a formal expression. If we have no information relevant to the actual value of a parameter, the probability must be chosen so as to express the fact that we have none. It must say nothing about the value of the parameter, except the bare fact that it may possibly, by its very nature, be restricted to lie within certain definite limits."

Jeffreys (1948, p. 168) has proved a remarkable theorem that when a sample is large, the results given by the theory of inverse probability are indistinguishable from those given by Fisher's method of maximum likelihood, even though we may assume any arbitrary distribution for prior probabilities of parameters. Thus, if the sample is large, it is the likelihoods of hypotheses which play a dominant part, and any arbitrariness in the assumed prior probabilities will make little difference to the results. In Jeffreys' theory of probability, the method of maximum likelihood is thus introduced, not as an independent postulate, but as a *derivative* principle, as a large-sample approximation to the principle of inverse probability.

An interesting analogy from athletics will be found appropriate in understanding the satisfactory results given by the principle of inverse probability in cases mentioned above. The different hypotheses can be compared to athletes in a running race. The prior probabilities are the "initial starts" given to the athletes, and the likelihoods are the "speeds" of the athletes. If all the athletes are given the same initial start, the fastest athlete will win. If all the athletes are equally fast, the athlete with longest initial start will win. In this case the speeds are of no consequence. The race is as if it is not run at all, being decided purely by the initial starts. Again, if the initial starts and the speeds both vary, but the distance of the race is sufficiently long (as in a marathon), the initial starts will be ignorable, and the race will be won by the fastest athlete.

1.5 Difficulties About Bayes' Dictum

We shall now show how indiscriminate use of Bayes' dictum leads to inconsistent results. Such a case arises when we have to estimate an unknown parameter with continuous range of values. For if α is such a parameter, and if we know nothing about the prior probability distribution of α, Bayes' dictum will give a uniform prior-probability distribution of α, i.e.,

$$P(d\alpha|H) \propto d\alpha. \tag{1-8}$$

Now suppose we make the transformation $\beta = \alpha^3$, and take β as the parameter. Since β is unknown, application of Bayes' dictum gives

$$P(d\beta|H) \propto d\beta. \tag{1-9}$$

But since $d\beta = 3\alpha^2\ d\alpha$, the prior-probability distribution of α, as deduced from Eq. (1-9), is

$$P(d\alpha|H) \propto \alpha^2\ d\alpha, \tag{1-10}$$

which contradicts Eq. (1-8).

Inconsistency will also arise, in general, if instead of Bayes' dictum, we take the same function or two different artibrary functions to represent the prior-probability distributions of α and β. Thus, if we use a function $f(\alpha_i)$ to represent the p.d.f. of the prior-probability distribution of the unknown parameters α_i, the parameters α_i are transformed to the parameters α'_j, and if we take the same function $f(\alpha'_j)$, or any other arbitrary function $\phi(\alpha'_j)$, for the p.d.f. of the prior probability of the new set of parameters α'_j, then the probabilities of the corresponding regions of the two sets of parameters will be different in general. This contradicts the requirement of consistency that all equivalent propositions have the same probability. (Jeffreys, 1948, p. 8, Rule 2.)

Even if we do not consider transformation of an unknown parameter, Jeffreys has given cases where uniform distribution of prior probability gives unsatisfactory results. One of these cases is when the range of possible values of an unknown parameter is limited at one end only (as in the case when the parameter is standard deviation or variance). Jeffreys has also shown that the principle of uniform distribution of prior probability would make impossible tests of significance for testing of hypotheses. An illustration of this occurs in the famour Bayes-Laplace theory of sampling from an infinite (or large) population where uniform distribution of prior probability has been shown by Broad (1918) to lead to the following unsatisfactory results: (1) A large homogeneous sample will give only a moderate probability that a further sample comparable in size to the first sample will be of the same type, though it will give a high probability that the next member will be of the same type. (2) Sampling will never give a high probability that the whole population is homogeneous unless the sample constitutes a large proportion of the whole population.

Laplace's rule of succession (which follows from the uniform assessment of prior probability) had been, for a long time, generally appealed to as a justification of induction. But Broad showed that it was no justification whatever for attaching even a moderate probability to a general rule if the possible instances of the rule are many times more numerous than those already investigated. If we are ever to attach a high probability to a general rule on any practical amount of evidence, it is necessary that it must have moderate probability to start with.

1.6 Hypothesis with Nonzero Prior Probability: Approach to Certainty of an Inductive Inference

Let $p_1, p_2, p_3, \ldots$ be a sequence of observable consequences of a hypothesis q with nonzero prior probability. Suppose further that all consequences up to p_n are verified. Jeffreys proves an interesting proposition,

$$P(p_{n+1}|p_1, \ldots p_n H) \to 1 \quad \text{as } n \to \infty. \tag{1-11}$$

The present writer (Huzurbazar, 1955c) has strengthened this proposition in the form

$$P(p_{n+1}p_{n+2} \cdots p_{n+m}|p_1 p_2 \cdots p_n H) \to 1, \tag{1-12}$$

when n and m tend to infinity in any manner whatsoever. We have by the product rule

$$P(p_1 p_2 \cdots p_n|H) = \frac{P(q|H)}{P(q|p_1 p_2 \cdots p_n H)}. \tag{1-13}$$

Write $U_n = P(p_1p_2 \cdots p_n|H)$. Then from Eq. (1-13), $U_n \geq P(q|H)$, a fixed positive number. Since

$$U_{n+1} = U_n\, P(p_{n+1}|p_1p_2 \cdots p_nH),$$

we have $U_{n+1} \leq U_n$. The sequence U_n is therefore monotone decreasing and tends to a limit $\ell \geq P(q|H) > 0$. Therefore

$$\frac{U_{n+m}}{U_n} \to \frac{\ell}{\ell} = 1,$$

when n and m tend to infinity in any manner whatsoever. Now

$$\frac{U_{n+m}}{U_n} = P(p_{n+1}p_{n+2} \cdots p_{n+m}|p_1p_2 \cdots p_nH),$$

so that we have Eq. (1-12).

In words, Eq. (1-12) means: "Repeated verifications of the consequences of a hypothesis with nonzero prior probability will make it almost certain that *any number* of further consequences of it will be verified." The important distinction between the propositions (1-11) and (1-12) is that whereas Eq. (1-11) ensures certainty of only the *next* consequence of the hypothesis, Eq. (1-12) ensures certainty of the set of *any number* of further consequences of the hypothesis. For justification of our confidence in inductive inferences, it is necessary to have a strong proposition like Eq. (1-12).

It is interesting to observe that when the conditions of the proposition are not satisfied [e.g., when $P(q|H)$ is zero or an infinitesimal] we may come across cases where we have Eq. (1-11) but not Eq. (1-12). An illustration of this is the law of succession in the Bayes-Laplace theory of sampling from an infinite (or large) population. If q is the hypothesis that the population is homogeneous in a property ϕ, and if in a sample of size n all members are found to be ϕ's, then the probability that the next member will be ϕ (on uniform assessment of prior probability) is $(n+1)/(n+2)$, and the probability that the next m members will be ϕ's is $(n+1)/(n+m+1)$. Here

$$P(p_{n+1}|p_1p_2 \ldots p_nH) = \frac{n+1}{n+2} \to 1 \quad \text{as } n \to \infty.$$

But

$$P(p_{n+1}p_{n+2} \ldots p_{n+m}|p_1p_2 \ldots p_nH) = \frac{n+1}{n+m+1}$$

does not tend to a definite limit as n and m tend to infinity in any manner. Thus the law of succession does not lead to justification of induction. The reason for the failure of the proposition (1-12) in this case is that $P(q|H)$ is an infinitesimal on the uniform assessment of prior probability. On the other hand, if the Bayes-Laplace theory is modified, as has been done by Jeffreys, by taking $P(q|H) = k$, where k is a fixed positive number, we get both Eqs. (1-11) and (1-12).

Another important point to be noted is that the conclusion in Eq. (1-12) does not contain q at all. It would still stand if q were false. It does not say that q approaches certainty, merely that inferences from it do so. Thus a well-verified hypothesis will probably continue to lead to correct inferences even if it may be wrong. Jeffreys (1948, p. 39) makes some very interesting remarks:

> Now in science one of our troubles is that the alternatives available for consideration are not always an exhaustive set. An unconsidered one may escape attention for centuries. The unconsidered hypothesis, if it had been thought of, would either (1) have led to the same consequences p_1, p_2, ... or (2) to different consequences at some stage. In the latter case the data would have been enough to dispose of it, and the fact that it was not thought of has done no harm. In the former case the considered and the unconsidered alternatives would have the same consequences, and will presumably continue to have the same consequences. The unconsidered alternative becomes important only when it is explicitly stated and a type of observation can be found where it would lead to different predictions from the old one. The rise into importance of the theory of general relativity is a case in point. Even though we now know that the systems of Euclid and Newton need modification, it was still legitimate to base inferences on them until we knew what particular modification was needed. The theory of probability makes it possible to respect the great men on whose shoulders we stand.

1.7 Some Fairly Satisfactory Prior-Probability Forms

Though the problem of the choice of the best assessment of prior probabilities has not been yet solved, Jeffreys (1948, Sec. 3.1) has considered in detail the problem of finding a way of saying that the magnitude of a parameter is unknown, when none of the possible values need special attention. He has given two interesting rules which appear to cover the commonest cases: (1) If the parameter may have any value in a finite range, or from $-\infty$ to ∞, its prior probability should be taken as uniformly distributed. (2) If the parameter may have any value from 0 to ∞, the prior probability of its logarithm should be taken as uniformly distributed.

Thus, if μ is the unknown location parameter with range $(-\infty, \infty)$, Rule 1 gives $P(d\mu|H) \propto d\mu$ for the prior-probability distribution of μ. If σ is the unknown standard deviation or scale parameter of a distribution, its range of possible values is $(0, \infty)$ and Rule 2 gives

$$P(d\sigma|H) \propto d(\log \sigma) = \frac{d\sigma}{\sigma},$$

for the prior-probability distribution of σ, Jeffreys has shown how this form is more satisfactory than the form $P(d\sigma|H) \propto d\sigma$ given by the uniform distribution of the prior probability of σ. Gauss also seems to have found something unsatisfactory about the assumption of uniform distribution of the prior probability of σ.

It is remarkable that assuming these prior-probability forms for the unknown parameters μ and σ of a normal distribution, Jeffreys obtains with great ease several inverse probability distributions of the parameters. A curious point to be noted is that the posterior-probability distribution of μ (σ known) is normal with mean $\bar{x}$ and standard deviation $\sigma/\sqrt{n}$, where $\bar{x}$ is the mean of a given sample of size n. The posterior-probability distribution of μ (σ unknown) follows Student's t-distribution. There is thus a remarkable agreement between the results given in the direct theory and those obtained by the method of inverse probability as developed by Jeffreys.

The prior-probability form $d\sigma/\sigma$ possesses the property of invariance under the transformation $\sigma' = \lambda_1 \sigma^{\lambda_2}$, where λ_1 and λ_2 are constants. For

$$\frac{d\sigma'}{\sigma'} = \lambda_2 \frac{d\sigma}{\sigma}.$$

Therefore

$$P(d\sigma' | H) \propto \frac{d\sigma'}{\sigma'}.$$

In particular, the form secures consistency if we estimate the variance σ^2 in place of the standard deviation σ, or if we estimate the precision constant of a normal distribution in place of its standard deviation.

It should be remarked, however, that the form does not possess the invariance property for any general transformation of the parameter.

2. Distributions Admitting Sufficient Statistics

2.1 Introduction

The purpose of this section is to give a brief account of the most general forms of distributions admitting sufficient statistics for the parameters. These forms will be used in the subsequent sections to develop a theory of invariants of distributions admitting sufficient statistics. The invariants obtained will be applied in stating the prior probability of parameters in estimation problems.

2.2 Sufficient Statistics

The notion of a sufficient statistic, due to Fisher (1922), plays an important role in statistics. The characteristic property of a sufficient statistic is that if T_1 is a sufficient statistic for the unknown parameter α of a distribution and if T_2 is any other statistic, then if T_1 is fixed, the conditional distribution of T_2 is independent of α. Consequently, when T_1 is known, knowledge of

T_2 gives no further information about α. A sufficient statistic extracts the whole of the relevant information available in a sample about the unknown parameter, whatever the size of the sample may be, and is spoken of as the best statistic when it exists, in Fisher's theory of estimation.

Sufficient statistics are shown to have some remarkable properties in samples of any size, in Fisher's theory of point estimation and his theory of fiducial probability. They also possess interesting properties in other theories of statistical estimation or tests of significance such as Jeffreys' theory of probability, Neyman's theory of confidence intervals, the Neyman-Pearson theory of testing of statistical hypotheses, and the Aitken-Silverstone theory of estimation by minimum variance. Curiously enough, the cases where results of utmost satisfaction are achieved in all these theories are usually found to be associated with the existence of sufficient statistics. Moreover, in cases where sufficient statistics exist, there is often agreement among the results of the different theories.

2.3 Distributions Admitting Sufficient Statistics

The existence of a sufficient statistic for the parameter of a distribution will ultimately depend upon the form of the distribution itself. Unless a distribution is of some special character, we cannot hope to obtain a sufficient statistic and we are therefore presented with the important question: What is the most general class of distributions which possess a sufficient statistic for a parameter? An answer to this question was obtained by Koopman (1936) and Pitman (1936). If $f(x, \alpha)$ is the p.d.f. of a distribution depending on an unknown parameter α and if $\partial^2 f/(\partial x\, \partial\alpha)$ exists, then it is necessary and sufficient for the existence of a sufficient statistic for α that $f(x, \alpha)$ be of the form

$$f(x, \alpha) = \exp[u(\alpha)\, v(x) + A(x) + B(\alpha)], \tag{2-1}$$

where u and B are functions of α only, and v and A are functions of x only.

It is easy to see that if a distribution is in form (2-1), and if $x_1, x_2, \ldots, x_n$ is a sample of n independent observations, then $T = \sum_{i=1}^{n} v(x_i)$ is a sufficient statistic for α.

It is interesting to note that Fisher (1934) had incidently come very near the form (2-1) though he did not state it explicitly in that form. More recently Jeffreys (1959) has extended the proof without assuming the existence of $\partial^2 f/(\partial x\, \partial\alpha)$ so as to cover the cases when either x or α or both may assume only discrete values.

It can be verified easily that the normal distribution is of the form (2-1), both as regards the parameters of the mean and the standard deviation separately. Both the Poisson distribution and the binomial distribution admit sufficient statistics and can be written in the form (2-1), even though both are discrete distributions. It is interesting to verify that the form (2-1) includes, in fact, all the known distributions possessing a sufficient statistic for a parameter.

Koopman has also considered the case of several unknown parameters. Let a distribution depend on the parameters α_j (j = 1, 2, ..., p). Then the most general form of distributions admitting a set of *jointly* sufficient statistics for the parameters α_j is

$$f(x, \alpha_j) = \exp\left[\sum_{k=1}^{p} u_k(\alpha_j)\, v_k(x) + A(x) + B(\alpha_j)\right], \qquad (2\text{-}2)$$

where the u_k and B are functions of the α_j's, and the v_k and A are functions of x.

If $x_1, x_2, \ldots, x_n$ is a sample of n members, and if $T_k = \sum_{i=1}^{n} v_k(x_i)$, it can be shown that the distribution of any other statistic will be independent of the parameters α_j when the T_k (k = 1, 2, ..., p) are fixed. The T_k constitute a set of jointly sufficient statistics for the α_j. A set of jointly sufficient statistics also will be spoken of as a "sufficient set of statistics". It does not follow that each T_k is sufficient for the corresponding α_j when the other parameters are known, in which case the statistics are "completely sufficient" (Kendall, 1946, p. 39). But such cases are extremely rare.

The form (2-2), as well as the form (2-1), holds good for multivariate distributions also where we should simply write x_q for x, the x_q being variates. It may be verified that Eq. (2-2) includes all the known distributions which admit jointly sufficiently statistics, those well known being the univariate normal distribution, the multivariate normal distribution, the Beta distribution, Pearson's Type II distribution with known terminus, and the multinomial distribution. (The numbering of the Pearson types followed herein is that of Elderton, 1938.) It is interesting to note that all the Pearson distributions have sufficient statistics for some of the parameters if the others are assumed known.

So far we have tacitly assumed that the range of the distribution is independent of the parameters to be estimated. Pitman has obtained the most general form of distributions admitting a sufficient statistic when the range depends on α as

$$f(x, \alpha) = \frac{g(x)}{h(\alpha)} \qquad [a(\alpha) \le x \le b(\alpha)], \tag{2-3}$$

where $a(\alpha)$ and $b(\alpha)$ are respectively either monotone-increasing and monotone-decreasing, or else monotone-decreasing and monotone-increasing functions of α. [Pitman, however, assumes for simplicity the range of the distribution to be from α to $b(\alpha)$, where $b(\alpha)$ is monotone decreasing. But the argument can be easily extended to the range $a(\alpha) \le x \le b(\alpha)$ with the restrictions we have stated.] In special cases one of the extremities a or b may be fixed.

Denote by a^{-1} and b^{-1} the functions inverse to a and b respectively, and X_1 and X_n the smallest and the greatest members respectively in a sample of n observations. Then if $\hat{\alpha}$ is equal to the smaller of $a^{-1}(X_1)$ and $b^{-1}(X_n)$ if $a(\alpha)$ increases and $b(\alpha)$ decreases, and equal to the greater of $a^{-1}(X_1)$ and $b^{-1}(X_n)$ if $a(\alpha)$ decreases and $b(\alpha)$ increases with α, it can be shown that when $\hat{\alpha}$ is fixed, the distribution of any other statistic is independent of α, so that $\hat{\alpha}$ is a sufficient statistic for α. As well-known illustrations of

Eq. (2-3) we have

1. The rectangular distribution:

$$f(x, \alpha) = \frac{1}{2\alpha} \quad (-\alpha \le x \le \alpha).$$

2. Pearson's type X distribution:

$$f(x, \alpha) = e^{-(x-\alpha)} \quad (\alpha \le x < \infty).$$

3. Pearson's type XI distribution:

$$f(x, \alpha) = (m-1)\alpha^{m-1}x^{-m} \quad (\alpha \le x < \infty);$$

($m > 1$ is a known constant).

2.4 Distributions Admitting an Augmented Sufficient Set of Statistics

When no sufficient statistic exists for a parameter α, it may be possible to find a set of ν *independent* statistics which together contain all the information available in a sample about the parameter α, ν being the smallest integer having this property. Pitman calls such a set "a sufficient set of statistics," but as this term is likely to be confused with a sufficient set of statistics for many parameters referred to in Sec. 2.3, we propose to designate the set in question by the term "an augmented sufficient set of statistics." In the case of many parameters α_j ($j = 1, 2, \ldots, p$), an augmented sufficient set of statistics is a set ν ($\nu > p$) independent statistics which together contain all the information in a sample about the parameters α_j. An augmented sufficient set of statistics may be regarded as a sort of extension of the notion of a sufficient set of statistics in which the number of statistics equals the number of parameters to be estimated. For a single parameter α Pitman has shown that the most general form of distributions admitting an augmented sufficient set of ν statistics is

$$f(x, \alpha) = \exp\left[\sum_{k=1}^{\nu} u_k(\alpha)\, \nu_k(x) + A(x) + B(\alpha)\right]. \tag{2-4}$$

If $x_1, x_2, \ldots, x_n$ is a sample of n members, and if $T_k = \sum_{i=1}^{n} v_k(x_i)$, the T_k $(k = 1, 2, \ldots, \nu)$ constitute an augmented sufficient set of ν statistics for the parameter α. The extension to the case of several parameters is on similar lines, and the most general form of distributions admitting an augmented sufficient set of ν statistics for the p parameters α_j is

$$f(x, \alpha_j) = \exp\left[\sum_{i=1}^{\nu} u_k(\alpha_j)\, v_k(x) + A(x) + B(\alpha_j)\right]. \tag{2-5}$$

As before, for multivariate distributions x should be replaced by x_q in Eqs. (2-4) and (2-5).

As an illustration of Eq. (2-4) consider for simplicity the bivariate normal distribution with zero (or known) means,

$$f(x, y, \sigma_1, \sigma_2, \rho) = \frac{1}{2\pi\sigma_1\sigma_2[1 - \rho^2]^{\frac{1}{2}}} \exp\left[-\frac{1}{2(1-\rho^2)}\left(\frac{x^2}{\sigma_1^2} - \frac{2\rho xy}{\sigma_1\sigma_2} + \frac{y^2}{\sigma_2^2}\right)\right].$$

When all the three parameters σ_1, σ_2, and ρ are unknown, there is a sufficient set of statistics for them; but if σ_1 and σ_2 are known, there is no sufficient statistic for ρ, the correlation coefficient. But when σ_1 and σ_2 are known, we can write

$$f(x, y, \rho) = \exp\left\{-\frac{1}{2(1-\rho^2)}\left(\frac{x^2}{\sigma_1^2} + \frac{y^2}{\sigma_2^2}\right) + \frac{\rho xy}{\sigma_1\sigma_2(1-\rho^2)} - \log\left[2\pi\sigma_1\sigma_2(1-\rho^2)^{\frac{1}{2}}\right]\right\}.$$

which is of the form (2-4) with

$$v_1(x, y) = \frac{x^2}{\sigma_1^2} + \frac{y^2}{\sigma_2^2} \quad \text{and} \quad v_2(x, y) = xy.$$

Then

$$T_1 = \sum_{i=1}^{n} v_1(x_i, y_i) \quad \text{and} \quad T_2 = \sum_{i=1}^{n} v_2(x_i, y_i)$$

constitute an augmented sufficient pair of statistics for ρ.

When the range depends on the parameter to be estimated, there are two types of distributions admitting an augmented sufficient set of statistics, as shown by Pitman. But the type with which we shall be concerned is

$$f(x, \alpha) = \frac{g(x)}{h(\alpha)} \quad [a(\alpha) \leq x \leq b(\alpha)], \tag{2-6}$$

where $a(\alpha)$ and $b(\alpha)$ are no longer subject to the monotonic restrictions of Eq. (2-3). There is now no sufficient statistic for α, but the smallest and the greatest members in a sample together contain all the information about α and hence they constitute an augmented sufficient pair of statistics for α.

Distributions (2-4) - (2-6) are a sort of extension of the distributions (2-1) - (2-3) respectively.

2.5 Properties of Distributions Admitting Sufficient Statistics

Distributions admitting sufficient statistics possess interesting mathematical and statistical properties. We shall take the forms (2-1) - (2-3) and (2-4) - (2-6) as fundamental forms. Starting from these forms the present writer (Huzurbazar, 1948, 1949a, 1949b, 1955a, 1955b, 1956) has obtained several interesting mathematical and statistical properties of distributions admitting sufficient statistics.

Though we shall be often concerned with distributions admitting sufficient statistics (Sec. 2.3), we shall also consider sometimes distributions admitting an augmented sufficient set of statistics (Sec. 2.4). It may be remarked here that the properties possessed by the latter class are not, in general, so numerous and interesting as those possessed by the former one.

2.6 Posterior Probability Distribution of Parameters when Sufficient Statistics Exist

A remarkable property proved by the author (Huzurbazar, 1948) is that for all distributions admitting sufficient statistics, the posterior-probability distribution of the parameters has the *same* form as the parent distribution itself.

For if a distribution admits a set of jointly sufficient statistics for the parameters α_j $(j = 1, 2, \ldots, p)$, it is in the form

$$f(x, \alpha_j) = \exp\left[\sum_{k=1}^{p} u_k(\alpha_j)\, v_k(x) + A(x) + B(\alpha_j)\right]. \qquad (2\text{-}7)$$

If $x_1, x_2, \ldots, x_n$ is a sample of n members from the distribution,

$$P(dx_1\, dx_2 \ldots dx_n | \alpha_j H) = \exp\left[\sum_{k=1}^{p} u_k(\alpha_j) \sum_{i=1}^{n} v_k(x_i) + \sum_{i=1}^{n} A(x_i) + n\, B(\alpha_j)\right] dx_1\, dx_2 \ldots dx_n$$

$$= \exp\left[\sum_{k=1}^{p} u_k(\alpha_j)\, v'_k(T_j) + \sum_{i=1}^{n} A(x_i) + n\, B(\alpha_j)\right] dx_1\, dx_2 \ldots dx_n,$$

where $\sum_{i=1}^{n} v_k(x_i) = v'_k(T_j)$ is a function of the p sufficient statistics T_j of the set. Let

$$P(d\alpha_1\, d\alpha_2 \ldots d\alpha_p | H) \propto \psi(\alpha_j)\, d\alpha_1\, d\alpha_2 \ldots d\alpha_p.$$

Then by the principle of inverse probability

$$P(d\alpha_1\, d\alpha_2 \ldots d\alpha_p | \theta H) \propto \psi(\alpha_j) \exp\left[\sum_{k=1}^{p} u_k(\alpha_j)\, v'_k(T_j) + \sum_{i=1}^{n} A(x_i) + n\, B(\alpha_j)\right] d\alpha_1\, d\alpha_2 \ldots d\alpha_p$$

$$\propto \exp\left[\sum_{k=1}^{p} u_k(\alpha_j)\, v'_k(T_j) + B'(\alpha_j)\right], \qquad (2\text{-}8)$$

where $\exp[B'(\alpha_j)] = \psi(\alpha_j) \exp[n\, B(\alpha_j)]$, and $\exp[\sum_{i=1}^{n} A(x_i)]$ is omitted, being independent of the α_j. If the constant of proportionality in Eq. (2-8) is $\exp[A'(T_j)]$, the p.d.f. of the posterior-probability distribution of the α_j is

$$\phi(\alpha_j, T_j) = \exp\left[\sum_{k=1}^{p} u_k(\alpha_j)\, v_k'(T_j) + A'(T_j) + B(\alpha_j)\right], \tag{2-9}$$

which has the same form as Eq. (2-7).

If the range of the distribution depends on the parameter α, we have

$$f(x, \alpha) = \frac{g(x)}{h(\alpha)} \quad [a(\alpha) \leq x \leq b(\alpha)]. \tag{2-10}$$

$$P(dx_1\, dx_2 \ldots dx_n | \alpha H) = \frac{g(x_1)\, g(x_2) \ldots g(x_n)}{[h(\alpha)]^n}\, dx_1\, dx_2 \ldots dx_n.$$

Let $P(d\alpha | H) \propto \psi(\alpha)\, d\alpha$. By the principle of inverse probability

$$P(d\alpha | \theta H) \propto \frac{\psi(\alpha)}{[h(\alpha)]^n}\, d\alpha,$$

where we have omitted $g(x_1)\, g(x_2) \ldots g(x_n)$, being independent of α.

Let

$$G(\alpha) = \frac{\psi(\alpha)}{[h(\alpha)]^n}.$$

Then $P(d\alpha | \theta H) \propto G(\alpha)\, d\alpha$. (2-11)

In this case, given the sample, the range of the possible values of α is a function of T, a sufficient statistic (which is a function of the two extreme observations in the sample). Let the range of the possible values of α, given T, be $A(T) \leq \alpha \leq B(T)$.

Hence if $\int_{A(T)}^{B(T)} G(\alpha)\, d\alpha = H(T)$,

from Eq. (2-11) the p.d.f. of the posterior-probability distribution of α is

$$\phi(\alpha, T) = \frac{G(\alpha)}{H(T)} \quad [A(T) \leq \alpha \leq B(T)], \tag{2-12}$$

which has the same form as Eq. (2-10).

3. Invariance Theory of Jeffreys

3.1 Invariance Rule

As we noted in Sec. 1.5, a major objection raised by the opponents of inverse probability is that use of arbitrary prior-probability forms in problems such as estimation of parameters leads to inconsistent results when the parameters undergo transformations. This is no doubt a serious objection. A given probability distribution can be expressed in different forms either by transforming the variates or the parameters or both. In such a case it is essential that the prior probability of parameters in different forms should conform to the requirement of consistency that all equivalent propositions have the same probability. The invariance theory stated by Jeffreys (1946, 1948) meets the objection in question by using invariants for stating the prior probability of parameters. (It will be assumed that when we consider transformations of both the variates and the parameters, transformations of the variates will not involve the parameters and vice-versa. This restriction is necessary if the transformations are to serve any useful purpose in problems of estimation or testing of hypotheses.)

An invariance rule for stating the prior probability of parameters of a distribution may be formally defined as a rule which will lead to the same prior probability for corresponding regions of the parameters when the parameters as well as the variates undergo any unknown transformations. An invariance rule ensures that any arbitrariness in the form in which a distribution is expressed makes no difference to the results.

But mere consistency is not enough. It is also desirable that an invariance rule should lead to results agreeing with common sense

(Jeffreys, 1948, p. 10, requirement 7). An invariance rule should be tested to determine whether it leads to sensible prior-probability forms for parameters in a given distribution.

3.2 Jeffreys' System of Invariants

Consider two probability distributions of a variate x with distribution functions P and P'. Then Jeffreys (1946) defines

$$I_m = \int |(dP')^{1/m} - (dP)^{1/m}|^m, \tag{3-1}$$

$$J = \int \log \frac{dP'}{dP}\, d(P' - P), \tag{3-2}$$

where the integrals are defined in the Stieltjes manner, by taking δP and $\delta P'$ for the same interval of x, forming the approximate sums, and then making the intervals of x tend to zero. The use of Stieltjes integrals covers both discrete and continuous distributions. Extension to multivariate distributions follows immediately. For continuous distributions with probability density functions p and p', Eqs. (3-1) and (3-2) become

$$I_m = \int |p'^{1/m} - p^{1/m}|^m\, dx, \tag{3-3}$$

$$J = \int (p' - p) \log \frac{p'}{p}\, dx. \tag{3-4}$$

For discrete distributions p and p' are simply the probabilities that x takes a particular value x, and the integrals in Eqs. (3-3) and (3-4) are replaced by summation over the set of values of x.

The expressions I_m and J have remarkable properties. They are all invariant for all nonsingular transformations of the variate and of the parameters of the distribution, and they are all positive-definite. They can therefore be regarded as providing measures of discrepancy or "distance" between two distributions.

The concept of distance in statistics was first introduced by Mahalanobis (1936), who used an invariant D^2 to define the distance between two multivariate normal populations. Since then a few more frequency theorists of probability have introduced some invariants for use in tests of significance and discrimination analysis. But the invariants defined by Jeffreys form a wider class of invariants of probability distributions. Kullback and Leibler (1951) have demonstrated that Mahalanobis' D^2 is equivalent to Jeffreys' J in the case in question. In a paper based on my suggestion, Adke (1958) has shown that the distance function Δ^2 of Bhattacharya (1942) is expressible in terms of the invariant I_2 of Jeffreys, and that most of the existing measures of divergence between two populations used by frequency theorists are derivable from the basic invariants I_2 and J.

Though I_m represents a set of invariants for different values of m, Jeffreys considers only I_2 on account of the highly complicated form of I_m $(m \neq 2)$.

An important case is that of two distributions having the *same* mathematical form but with *different* sets of values of the parameters. If the corresponding parameters in the two sets differ by infinitesimals, we get the differential forms of the invariants I_m and J.

The differential forms I_2 and J can be obtained as follows:

Let p (the p.d.f.) depend on the parameters α_i $(i = 1, 2, \ldots, n)$, and p' be obtained from p by changing α_i to $\alpha_i + d\alpha_i$, so that $p' = p + dp$. Then from Eq. (3-3), to order $d\alpha_i\, d\alpha_k$, and using the summation convention with respect to the suffixes i, k, we have

$$
\begin{aligned}
I_2 &= \int |p'^{1/2} - p^{1/2}|^2 \, dx \\
&= \int p\left[\left(\frac{p'}{p}\right)^{1/2} - 1\right]^2 dx \\
&= \int p\left[\left(1 + \frac{dp}{p}\right)^{1/2} - 1\right]^2 dx \\
&= \int p\left[\frac{dp}{2p}\right]^2 dx \\
&= \frac{1}{4}\int \frac{1}{p}\left[\frac{\partial p}{\partial \alpha_i}\, d\alpha_i\right]^2 dx \\
&= \frac{1}{4}\int \frac{1}{p}\frac{\partial p}{\partial \alpha_i}\frac{\partial p}{\partial \alpha_k}\, d\alpha_i\, d\alpha_k\, dx
\end{aligned}
$$

$$= \frac{1}{4} d\alpha_i \, d\alpha_k \int \frac{1}{p} \frac{\partial p}{\partial \alpha_i} \frac{\partial p}{\partial \alpha_k} \, dx.$$

Hence $I_2 = \frac{1}{4} g_{ik} \, d\alpha_i \, d\alpha_k$, (3-5)

where $g_{ik} = \int \frac{1}{p} \frac{\partial p}{\partial \alpha_i} \frac{\partial p}{\partial \alpha_k} \, dx.$ (3-6)

From Eq. (3-4), to order $d\alpha_i \, d\alpha_k$, we have

$$J = \int dp \log\left(1 + \frac{dp}{p}\right) dx$$

$$= \int \frac{(dp)^2}{p} \, dx$$

$$= \int \frac{1}{p}\left(\frac{\partial p}{\partial \alpha_i} \, d\alpha_i\right)^2 dx$$

$$= \int \frac{1}{p} \frac{\partial p}{\partial \alpha_i} \frac{\partial p}{\partial \alpha_k} \, d\alpha_i \, d\alpha_k \, dx$$

$$= d\alpha_i \, d\alpha_k \int \frac{1}{p} \frac{\partial p}{\partial \alpha_i} \frac{\partial p}{\partial \alpha_k} \, dx.$$

Hence $J = g_{ik} \, d\alpha_i \, d\alpha_k$. (3-7)

Thus J and $4I_2$ have the same quadratic differential form, which again has the form of the square of an element of distance of curvilinear coordinates.

If we transform to any other set of parameters α'_j, I_2 and J are unaltered, and

$$J = g'_{j\ell} \, d\alpha'_j \, d\alpha'_\ell, \tag{3-8}$$

where $g'_{j\ell} = g_{ik} \frac{\partial \alpha_i}{\partial \alpha'_j} \frac{\partial \alpha_k}{\partial \alpha'_\ell}$, (3-9)

Then $|g'_{j\ell}| = |g_{ik}| \left|\frac{\partial \alpha_i}{\partial \alpha'_j}\right| \left|\frac{\partial \alpha_k}{\partial \alpha'_\ell}\right|$,

or $|g'_{j\ell}| = |g_{ik}| \left|\frac{\partial \alpha_i}{\partial \alpha'_j}\right|^2$. (3-10)

Now in the transformation of a multiple integral,

$$d\alpha_1\ d\alpha_2\ \ldots\ d\alpha_n = \left|\left|\frac{\partial\alpha_i}{\partial\alpha'_j}\right|\right|\ d\alpha'_1\ d\alpha'_2\ \ldots\ d\alpha'_n. \tag{3-11}$$

From Eqs. (3-10) and (3-11),

$$d\alpha_1\ d\alpha_2\ \ldots\ d\alpha_n = \frac{|g'_{j\ell}|^{1/2}}{|g_{ik}|^{1/2}}\ d\alpha'_1\ d\alpha'_2\ \ldots\ d\alpha'_n,$$

or

$$|g_{ik}|^{1/2}\ d\alpha_1\ d\alpha_2\ \ldots\ d\alpha_n = |g'_{j\ell}|^{1/2}\ d\alpha'_1\ d\alpha'_2\ \ldots\ d\alpha'_n. \tag{3-12}$$

Hence I_2 and J yield the important differential invariant $|g_{ik}|^{1/2}\ d\alpha_1\ d\alpha_2\ \ldots\ d\alpha_n$. It is invariant for all nonsingular transformations of the parameters as well as the variates.

3.3 Jeffreys' Invariance Rule

The differential invariant obtained in the previous section forms the basis of the following invariance rule given by Jeffreys for stating the prior probability of parameters in estimation problems: "Take the prior-probability density of parameters to be proportional to $|g_{ik}|^{1/2}$, where $|g_{ik}|$ is the determinant of the quadratic differential form of the invariant J."

This rule is applicable to all probability distributions which are differentiable with respect to parameters in it. The rule thus enables us to construct a consistent theory of probability for this wide class of distributions.

Other invariance rules were also proposed by Diananda (1946), and for a single parameter by Perks (1947), by consideration of the expressions for the variance or the variance-covariance matrix of the maximum-likelihood estimators of parameters in large samples. But those rules, though arrived at differently, are essentially particular cases of the rule given by Jeffreys, there being a relation between maximum likelihood and the invariance theory of Jeffreys (1948, p. 169).

3.4 Review of Some Results Obtained with Jeffreys' Invariance Rule

Jeffreys has considered application of his rule to several well-known distributions.

For the univariate normal distribution with mean α and standard deviation σ, the quadratic differential forms of I_2 and J are

$$4I_2 = J = \left(\frac{d\alpha}{\sigma}\right)^2 + 2\left(\frac{d\sigma}{\sigma}\right)^2 . \tag{3-13}$$

Three cases arise. If α is unknown but σ is known, the coefficient of $(d\alpha)^2$ is constant, giving a uniform prior-probability distribution for α over the range permitted. This agrees with the rule for satisfactory prior-probability form for a location parameter (Sec. 1.7).

If α is known but σ is unknown we have $|g_{ik}|^{1/2}\, d\sigma \propto d\sigma/\sigma$, so that the invariance rule leads to the appropriate prior-probability form for a standard deviation (or a scale parameter) mentioned in Sec. 1.7.

If α and σ are both unknown, $|g_{ik}|^{1/2}\, d\alpha\, d\sigma \propto d\alpha\, d\sigma/\sigma^2$. The invariance rule therefore leads to the form $d\alpha\, d\sigma/\sigma^2$ instead of the usual form $d\alpha\, d\sigma/\sigma$. This is unsatisfactory. In the usual situation in an estimation problem, α and σ are each capable of any value over a considerable range, and neither gives any appreciable information about the other. The prior-probability distributions of α and σ therefore should be independent. There is no trouble for α alone or σ alone; it arises only when they are considered both at once.

The same trouble arises when the location and scale parameters are considered together in the bivariate or multivariate normal distribution. An extra factor $1/\sigma$ would appear for each mean.

Consider a distribution such as that of partial correlation

$$P(dx_1\, dx_2 \ldots dx_n | a_{ik}, \sigma_i, H) = A \exp\left(-\frac{1}{2} W\right) dx_1\, dx_2 \ldots dx_n$$

where $W = \Sigma\Sigma\, a_{ik}x_i x_k/\sigma_i\sigma_k$, and the x_i are a set of observables. Here for each x_i there is a corresponding scale parameter σ_i, and the a_{ik} are numerical coefficients. $|g_{ik}|$ will be of the form $\pi_i B/\sigma_i^2$, where B is a numerical factor depending on the a_{ik}. The invariance rule leads to

$$P(d\sigma_i\ da_{k\ell}|H) \propto \prod_i \frac{d\sigma_i}{\sigma_i} B^{1/2} \prod da_{k\ell}. \tag{3-14}$$

In this case there is no difficulty in the introduction of any number of scale parameters.

Jeffreys therefore suggests a modification of the invariance rule so that we can deal with location parameters on the hypothesis that the scale and numerical parameters are irrelevent to them by simply taking their prior probability as being uniform. If α and σ are location and scale parameters in general, and a_i the numerical parameters, we can take

$$P(d\alpha\ d\sigma \prod da_i|H) \propto d\alpha\ |g_{ik}|^{1/2}\ d\sigma \prod da_i, \tag{3-15}$$

where $|g_{ik}|$ is found by varying only σ and the a_i, and is equal to $1/\sigma^2$ times a function of the a_i. This is invariant for transformations of the form

$$\alpha' = \alpha + \sigma\ f(a_i) \tag{3-16}$$

which is the only form of transformation of α that we should wish to make.

When α is the parameter of the binomial distribution, or is itself a chance, the invariance rule leads to the prior-probability form

$$P(d\alpha|H) = \frac{1}{\pi} \frac{d\alpha}{[\alpha(1-\alpha)]^{1/2}}. \tag{3-17}$$

Though this form may be recommended in some sitautions, the uniform distribution of prior probability for α is ordinarily satisfactory in estimation problems, besides being in conformity with the rule (Sec. 1.7) for a finite range, the range of α being (0, 1).

If α is the parameter of the Poisson distribution, Jeffreys has shown that the prior-probability form $d\alpha/\alpha$ is satisfactory in estimation problems. It is also in agreement with Rule 2 of Sec. 1.7, the range of α being $(0, \infty)$. But the invariance rule leads to the prior-probability form $d\alpha/\sqrt{\alpha}$.

When the parameter is a correlation coefficient, its range is finite, and uniform distribution of the prior probability is satisfactory. But the invariance rule does not lead to uniform prior-probability distribution of the correlation coefficient in a bivariate normal population.

The results we have mentioned show that the invariance rule given by Jeffreys, though it ensures consistency, unfortunately does not lead to the appropriate prior-probability forms in many cases.

The rule is not applicable to distributions which are not differentiable with regard to all the parameters in them, though a modification of the rule is suggested in some situations. The rule does not work for distributions when the range depends on the parameters to be estimated, such as the rectangular distribution.

Jeffreys (1948, p. 167) himself realizes the inadequacy of his invariance theory and has stressed the need for further research. He has also given (Jeffreys, 1954), an instructive resume of the invariance theory wherein he remarks: "The idea of invariance therefore goes a long way towards a general quantitative theory. I doubt myself whether it has yet been stated in the best possible form. For one thing, in some simple applications simpler methods lead directly to satisfactory results, whereas the methods based on J need some rather artificial-looking modifications. For another, in some problems the determinant $|g_{ik}|$ does not exist. But other invariants exist, and I think that fuller investigation of the possibilities of invariance theory would be worth the attention of workers in abstract spaces and differential geometry."

4. Some Properties of Jeffreys' Invariants

4.1 Introduction

Jeffreys has obtained the exact forms of the invariants I_2 and J for the univariate and bivariate normal distributions, the Poisson distribution, and the binomial and multinomial distributions; and for these distributions the exact forms of I_2 and J come out as *explicit* functions of the parameters. It is interesting to note that these distributions admit sufficient statistics for the parameters. In this section we are going to prove a remarkable theorem (Huzurbazar, 1955b) that for all distributions admitting sufficient statistics the exact forms of I_2 and J can be obtained as explicit functions of parameters of the distributions. The property holds good for any I_m if m is even. As mentioned in Sec. 3.2, Jeffreys has not investigated the properties of I_m $(m \neq 2)$ on account of its highly complicated form. Some properties of I_1 investigated by the present author (Huzurbazar, 1955d) will also be considered.

4.2 One Useful Integral

As given in Eq. (2-2), the most general form of distributions admitting sufficient statistics for the parameters α_j $(j = 1, 2, \ldots, p)$ is

$$f(x, \alpha_j) = \exp\left[\sum_{k=1}^{p} u_k(\alpha_j)\, v_k(x) + A(x) + B(\alpha_j)\right]. \tag{4-1}$$

Since $\int f(x, \alpha_j)\, dx \equiv 1$ for all admissible values of α_j, we have

$$\int \exp\left[\sum_{k=1}^{p} u_k(\alpha_j)\, v_k(x) + A(x)\right] dx = \exp[-B(\alpha_j)]. \tag{4-2}$$

Now the $u_k(\alpha_j)$ are p independent functions of the p independent parameters α_j. [The independence of the functions $u_k(\alpha_j)$ follows from the fact that the u_k can also be chosen as parameters of the distribution in place of the α_j. If the u_k are not independent, the number of independent parameters of the distribution can be reduced.

This will be contrary to the assumption that the α_j are a set of p independent parameters of the distribution.] We can express the α_j inversely as functions of the u_k's. Then $B(\alpha_j)$ can be expressed in terms of the u_k's as

$$B(\alpha_j) = b(u_k). \tag{4-3}$$

Then Eq. (4-2) becomes

$$\int \exp\left[\sum_{k=1}^{p} u_k(\alpha_j)\, v_k(x) + A(x)\right] dx = \exp[-b(u_k)]. \tag{4-4}$$

4.3 Exact Form of I_2

Let
$$f(x, \alpha_j) = \exp\left[\sum_{k=1}^{p} u_k(\alpha_j)\, v_k(x) + A(x) + B(\alpha_j)\right], \tag{4-5}$$

and
$$f(x, \alpha_j') = \exp\left[\sum_{k=1}^{p} u_k(\alpha_j')\, v_k(x) + A(x) + B(\alpha_j')\right], \tag{4-6}$$

so that $f(x, \alpha_j)$ and $f(x, \alpha_j')$ have the same mathematical form but with different sets of values of the parameters. We have

$$\begin{aligned} I_2 &= \int \{[f(x, \alpha_j')]^{1/2} - [f(x, \alpha_j)]^{1/2}\}^2\, dx \\ &= 2 - 2\int [f(x, \alpha_j)\, f(x, \alpha_j')]^{1/2}\, dx \\ &= 2 - 2\int \exp\left[\sum_{k=1}^{p}\left[\frac{1}{2}u_k(\alpha_j) + \frac{1}{2}u_k(\alpha_j')\right] v_k(x)\right. \\ &\qquad \left. + A(x) + \frac{1}{2}B(\alpha_j) + \frac{1}{2}B(\alpha_j')\right] dx \\ &= 2 - 2\exp\left[\frac{1}{2}B(\alpha_j) + \frac{1}{2}B(\alpha_j')\right]\int \exp\left[\sum_{k=1}^{p}\left[\frac{1}{2}u_k(\alpha_j)\right.\right. \\ &\qquad \left.\left. + \frac{1}{2}u_k(\alpha_j')\right] v_k(x) + A(x)\right] dx. \end{aligned} \tag{4-7}$$

Writing $\frac{1}{2}u_k(\alpha_j) + \frac{1}{2}u_k(\alpha_j')$ for u_k in Eq. (4-4), we have

$$\int \exp\left\{\sum_{k=1}^{p}\left[\frac{1}{2}u_k(\alpha_j) + \frac{1}{2}u_k(\alpha_j')\right]v_k(x) + A(x)\right\} dx$$

$$= \exp\left\{- b\left[\frac{1}{2}u_k(\alpha_j) + \frac{1}{2}u_k(\alpha_j')\right]\right\} \tag{4-8}$$

From Eqs. (4-7) and (4-8) we have

$$I_2 = 2 - 2\left\{\exp \frac{1}{2}B(\alpha_j) + \frac{1}{2}B(\alpha_j') - b\left[\frac{1}{2}u_k(\alpha_j) + \frac{1}{2}u_k(\alpha_j')\right]\right\}. \tag{4-9}$$

The curious point to be noted is that the function $A(x)$ remains unaltered in the integral on the left-hand side of Eq. (4-8), which enables us to evaluate that integral explicitly in virtue of Eq. (4-4).

Example. Consider the Type III distribution

$$f(x, a, p) = \frac{a^p e^{-ax} x^{p-1}}{\Gamma(p)} \quad (0 \le x < \infty).$$

We write

$$f(x, a, p) = \exp[-ax + p \log x - \log x + p \log a - \log \Gamma(p)].$$

Here $u_1 = a$, $u_2 = p$, and $B = p \log a - \log \Gamma(p)$. Expressing B in terms of u_1 and u_2,

$$B = u_2 \log u_1 - \log \Gamma(u_2) = b(u_1, u_2).$$

$$b\left(\frac{a + a'}{2}, \frac{p + p'}{2}\right) = \frac{p + p'}{2} \log \frac{a + a'}{2} - \log \Gamma\left(\frac{p + p'}{2}\right).$$

$$I_2 = 2 - 2 \exp\left[\frac{1}{2} p \log a - \frac{1}{2} \log \Gamma(p) + \frac{1}{2} p' \log a' - \frac{1}{2} \log \Gamma(p') - \frac{p + p'}{2} \log \frac{a + a'}{2} - \log \Gamma\left(\frac{p + p'}{2}\right)\right]$$

$$= 2 - \frac{2\,\Gamma[(p + p')/2]}{[\Gamma(p)\,\Gamma(p')]^{1/2}} \frac{a^{(1/2)p}\, a'^{(1/2)p'}}{[(a + a')/2]^{(1/2)(p + p')}}.$$

4.4 Exact Form of J

$$J = \int [f(x, \alpha_j') - f(x, \alpha_j)][\log f(x, \alpha_j') - \log f(x, \alpha_j)]\, dx$$

$$= \int [f(x, \alpha_j') - f(x, \alpha_j)]\left\{\sum_{k=1}^{p} [u_k(\alpha_j') - u_k(\alpha_j)]\, v_k(x) + B(\alpha_j') - B(\alpha_j)\right\} dx$$

$$= \sum_{k=1}^{p} [u_k(\alpha_j') - u_k(\alpha_j)]\left[\int v_k(x)\, f(x, \alpha_j')\, dx - \int v_k(x)\, f(x, \alpha_j)\, dx\right]$$

$$= \sum_{k=1}^{p} [u_k(\alpha_j') - u_k(\alpha_j)][E'\, v_k(x) - E\, v_k(x)],$$

where $E'\, v_k(x)$ and $E\, v_k(x)$ denote the expectations of $v_k(x)$ when the parameters are α_j' and α_j respectively. For brevity write

$$E_k' = E'\, v_k(x), \quad E_k = E\, v_k(x).$$

Then $$J = \sum_{k=1}^{p} [u_k(\alpha_j') - u_k(\alpha_j)](E_k' - E_k). \tag{4-10}$$

E_k' and E_k can be obtained as follows. We have

$$\frac{\partial}{\partial \alpha_r} \log f(x, \alpha_j) = \sum_{k=1}^{p} \frac{\partial u_k}{\partial \alpha_r} v_k(x) + \frac{\partial B}{\partial \alpha_r}.$$

Since $E\left[\frac{\partial}{\partial \alpha_r} \log f(x, \alpha_j)\right] \equiv 0,$

we have $$\sum_{k=1}^{p} \frac{\partial u_k}{\partial \alpha_r} E_k + \frac{\partial B}{\partial \alpha_r} = 0. \tag{4-11}$$

Setting $r = 1, 2, \ldots, p$ in Eq. (4-11), we have p simultaneous linear equations to determine the E_k. The E_k' are then obtained by writing α_j' for α_j in E_k. Thus Eqs. (4-10) and (4-11) enable us to express J explicitly in terms of the parameters.

The foregoing formulae are greatly simplified if we take the u_k as parameters and express $B(\alpha_j)$ in terms of them. Then Eq. (4-11) becomes simply

$$E_r + \frac{\partial B}{\partial u_r} = 0, \quad \text{so that} \quad E_r = -\frac{\partial B}{\partial u_r}.$$

From Eq. (4-10)

$$J = \sum_{k=1}^{p} (u_k' - u_k)\left(-\frac{\partial B'}{\partial u_k'} + \frac{\partial B}{\partial u_k}\right). \tag{4-12}$$

Writing $u_k' - u_k = du_k$, the differential form of J is

$$J = -\sum_{k=1}^{p}\sum_{\ell=1}^{p} \frac{\partial^2 B}{\partial u_k\, \partial u} du_k\, du_\ell. \tag{4-13}$$

For a single parameter, Eq. (4-13) becomes

$$J = -\frac{\partial^2 B}{\partial u^2} du^2. \tag{4-14}$$

4.5 Exact Form of I_m (m even)

For brevity we write $f(x, \alpha_j') = f'$, $f(x, \alpha_j) = f$, $u_k(\alpha_j') = u_k'$, $B(\alpha_j') = B'$, etc. Now

$$\begin{aligned} I_m &= \int |f'^{1/m} - f^{1/m}|^m\, dx \\ &= \int (f'^{1/m} - f^{1/m})^m\, dx \quad \text{(since m is even)} \\ &= \int \left[\sum_{r=0}^{m} (-1)^r \binom{m}{r} f'^{(m-r)/m} f^{r/m}\right] dx \\ &= \sum_{r=0}^{m} (-1)^r \binom{m}{r} \lambda_r, \end{aligned} \tag{4-15}$$

where

$$\begin{aligned} \lambda_r &= \int f'^{(m-r)/m} f^{r/m}\, dx \\ &= \int \exp\left[\sum_{k=1}^{p} \left(\frac{m-r}{m} u_k' + \frac{r}{m} u_k\right) v_k(x) + A(x) + \frac{m-r}{m} B' + \frac{r}{m} B\right] dx \end{aligned}$$

$$= \exp\left(\frac{m-r}{m} B' + \frac{r}{m} B\right) \int \exp\left[\sum_{k=1}^{p} \left(\frac{m-r}{m} u_k' + \frac{r}{m} u_k\right) v_k(x) + A(x)\right] dx. \tag{4-16}$$

Writing $[(m-r)/m]u_k' + (r/m)u_k$ for u_k in Eq. (4-4), we have

$$\int \exp\left[\sum_{k=1}^{p} \left(\frac{m-r}{m} u_k' + \frac{r}{m} u_k\right) v_k(x) + A(x)\right] dx = \exp\left[-b\left(\frac{m-r}{m} u_k' + \frac{r}{m} u_k\right)\right]. \tag{4-17}$$

From Eqs. (4-16) and (4-17)

$$\lambda_r = \exp\left[\frac{m-r}{m} B' + \frac{r}{m} B - b\left(\frac{m-r}{m} u_k' + \frac{r}{m} u_k\right)\right].$$

Then from Eq. (4-15)

$$I_m = \sum_{r=0}^{m} (-1)^r \binom{m}{r} \exp\left[\frac{m-r}{m} B' + \frac{r}{m} B - b\left(\frac{m-r}{m} u_k' + \frac{r}{m} u_k\right)\right]. \tag{4-18}$$

Curiously again, the function $A(x)$ remains unaltered in the integral in Eq. (4-17), which enables us to evaluate that integral explicitly in virtue of Eq. (4-4).

4.6 Some Properties of the Invariant I_1

The invariant $I_1 = \int |p' - p|\, dx$ seems to be useful in certain cases in stating the prior probability of a single parameter. Its drawback, however, is that it does not lead to the joint prior probability of two or more parameters. We shall now discuss some of the cases in which I_1 leads to the usual prior-probability forms.

4.6.1 The normal distribution

$$p = \frac{1}{\sigma\sqrt{2\pi}} \exp\left[-\frac{(x-\alpha)^2}{2\sigma^2}\right].$$

a. Suppose that σ is known but α is unknown. First we shall evaluate the differential form of I_1 for small variations of α. We have, to order $d\alpha$,

$$I_1 = d\alpha \int_{-\infty}^{\infty} \left|\frac{\partial p}{\partial \alpha}\right| dx.$$

$$\frac{\partial p}{\partial \alpha} = \frac{x - \alpha}{\sigma^3 \sqrt{2\pi}} \exp\left[- \frac{(x-\alpha)^2}{2\sigma^2}\right]$$

$$\frac{\partial p}{\partial \alpha} \gtrless 0, \quad \text{according as} \quad x \gtrless \alpha.$$

$$\text{Hence} \int_{-\infty}^{\infty} \left|\frac{\partial p}{\partial \alpha}\right| dx = \frac{1}{\sigma^3 \sqrt{2\pi}} \int_{-\infty}^{\alpha} (\alpha - x) \exp\left[- \frac{(x-\alpha)^2}{2\sigma^2}\right] dx$$

$$+ \frac{1}{\sigma^3 \sqrt{2\pi}} \int_{\alpha}^{\infty} (x - \alpha) \exp\left[- \frac{(x-\alpha)^2}{2\sigma^2}\right] dx$$

$$= \frac{2}{\sigma\sqrt{2\pi}}.$$

$$\text{Thus } I_1 = \frac{2}{\sigma\sqrt{2\pi}} d\alpha \propto d\alpha. \tag{4-19}$$

I_1 gives the appropriate prior-probability form of α when σ is known.

The exact form of I_1 in this case can be obtained as follows.

$$I_1 = \int_{-\infty}^{\infty} \left|\frac{1}{\sigma\sqrt{2\pi}} \exp\left[- \frac{(x-\alpha')^2}{2\sigma^2}\right] - \frac{1}{\sigma\sqrt{2\pi}} \exp\left[- \frac{(x-\alpha)^2}{2\sigma^2}\right]\right| dx$$

Here $p' > p$ if, and only if, $(x - \alpha')^2 \quad (x - \alpha)^2$, i.e., if

$$x > \frac{\alpha + \alpha'}{2} \quad \text{if} \quad \alpha' > \alpha,$$

$$\text{and} \quad x < \frac{\alpha + \alpha'}{2} \quad \text{if} \quad \alpha' < \alpha.$$

First let $\alpha' > \alpha$. Then

$$I_1 = \int_{-\infty}^{(\alpha+\alpha')/2} \left\{\frac{1}{\sigma\sqrt{2\pi}} \exp\left[- \frac{(x-\alpha)^2}{2\sigma^2}\right] - \frac{1}{\sigma\sqrt{2\pi}} \exp\left[- \frac{(x-\alpha')^2}{2\sigma^2}\right]\right\} dx$$

$$+ \int_{(\alpha+\alpha')/2}^{\infty} \left\{\frac{1}{\sigma\sqrt{2\pi}} \exp\left[- \frac{(x-\alpha')^2}{2\sigma^2}\right] - \frac{1}{\sigma\sqrt{2\pi}} \exp\left[- \frac{(x-\alpha)^2}{2\sigma^2}\right]\right\} dx$$

$$= \frac{4}{\sqrt{2\pi}} \int_0^{(\alpha'-\alpha)/2\sigma} e^{-u^2/2}\, du. \tag{4-20}$$

Similarly, if $\alpha' < \alpha$, we obtain

$$I_1 = \frac{4}{\sqrt{2\pi}} \int_0^{(\alpha-\alpha')/2\sigma} e^{-u^2/2}\, du. \tag{4-21}$$

Combining Eqs. (4-20) and (4-21),

$$I_1 = \frac{4}{\sqrt{2\pi}} \int_0^{|\alpha-\alpha'|/2\sigma} e^{-u^2/2}\, du. \tag{4-22}$$

We may also write Eq. (4-22) in terms of the "error function" as

$$I_1 = \left(\frac{2}{\pi}\right)^{1/2} \operatorname{erf}\left(\frac{|\alpha-\alpha'|}{2\sqrt{2}\,\sigma}\right). \tag{4-23}$$

The differential form (4-19) of I_1 can be deduced from Eq. (4-23) by writing $\alpha' - \alpha = d\alpha$.

b. Suppose now that α is known but σ unknown. First let us obtain the differential form of I_1 for small variations of σ. We have, to order $d\sigma$,

$$I_1 = d\sigma \int_{-\infty}^{\infty} \left|\frac{\partial p}{\partial \sigma}\right| dx.$$

$$\frac{\partial p}{\partial \sigma} = -\frac{1}{\sigma^2\sqrt{2\pi}} \exp\left[-\frac{(x-\alpha)^2}{2\sigma^2}\right] + \frac{(x-\alpha)^2}{\sigma^4\sqrt{2\pi}} \exp\left[-\frac{(x-\alpha)^2}{2\sigma^2}\right].$$

$\partial p/\partial\sigma > 0$ if, and only if, $|x - \alpha| > \sigma$. Hence

$$\int_{-\infty}^{\infty} \left|\frac{\partial p}{\partial \sigma}\right| dx = \int_{-\infty}^{\alpha-\sigma} \frac{\partial p}{\partial \sigma}\, dx - \int_{\alpha-\sigma}^{\alpha+\sigma} \frac{\partial p}{\partial \sigma}\, dx + \int_{\alpha+\sigma}^{\infty} \frac{\partial p}{\partial \sigma}\, dx$$

$$= \int_{-\infty}^{\infty} \frac{\partial p}{\partial \sigma}\, dx - 2\int_{\alpha-\sigma}^{\alpha+\sigma} \frac{\partial p}{\partial \sigma}\, dx$$

$$= -2\int_{\alpha-\sigma}^{\alpha+\sigma} \frac{\partial p}{\partial \sigma}\, dx \quad \left(\text{since} \int_{-\infty}^{\infty} \frac{\partial p}{\partial \sigma}\, dx \equiv 0\right)$$

$$= 2\int_{\alpha-\sigma}^{\alpha+\sigma} \frac{1}{\sigma^2\sqrt{2\pi}} \exp\left[-\frac{(x-\alpha)^2}{2\sigma^2}\right] dx - 2\int_{\alpha-\sigma}^{\alpha+\sigma} \frac{(x-\alpha)^2}{\sigma^4\sqrt{2\pi}} \exp\left[-\frac{(x-\alpha)^2}{2\sigma^2}\right] dx$$

$$= \frac{4}{\sigma\sqrt{2\pi e}}.$$

Therefore $I_1 = \frac{4}{\sqrt{2\pi e}} \frac{d\sigma}{\sigma} \propto \frac{d\sigma}{\sigma}$. (4-24)

Thus I_1 leads to the appropriate prior-probability form for σ when α is known.

The exact form of I_1 in this case can be obtained as follows.

$$I_1 = \int_{-\infty}^{\infty} \left| \frac{1}{\sigma'\sqrt{2\pi}} \exp\left[-\frac{(x-\alpha)^2}{2\sigma'^2}\right] - \frac{1}{\sigma\sqrt{2\pi}} \exp\left[-\frac{(x-\alpha)^2}{2\sigma^2}\right]\right| dx$$

Now $p' > p$ if, and only if,

$$\frac{(x-\alpha)^2}{2}\left(\frac{1}{\sigma^2} - \frac{1}{\sigma'^2}\right) > \log\frac{\sigma'}{\sigma},$$

i.e., if $(x-\alpha)^2 > \frac{2\log(\sigma'/\sigma)}{(1/\sigma^2) - (1/\sigma'^2)}$ if $\sigma' > \sigma$,

and $(x-\alpha)^2 < \frac{2\log(\sigma'/\sigma)}{(1/\sigma^2) - (1/\sigma'^2)}$ if $\sigma' < \sigma$.

Write $\rho^2 = \frac{2\log(\sigma'/\sigma)}{(1/\sigma^2) - (1/\sigma'^2)}$. (4-25)

First suppose that $\sigma' > \sigma$ so that $p' > p$, if, and only if, $|x-\alpha| > \rho$. Then

$$I_1 = \int_{-\infty}^{\alpha-\rho} (p'-p)\, dx + \int_{\alpha-\rho}^{\alpha+\rho} (p-p')\, dx + \int_{\alpha+\rho}^{\infty} (p'-p)\, dx$$

$$= 2\int_{\alpha-\rho}^{\alpha+\rho} (p-p')\, dx$$

$$= 2\int_{\alpha-\rho}^{\alpha+\rho} \frac{1}{\sigma\sqrt{2\pi}} \exp\left[-\frac{(x-\alpha)^2}{2\sigma^2}\right] dx - 2\int_{\alpha-\rho}^{\alpha+\rho} \frac{1}{\sigma'\sqrt{2\pi}} \exp\left[-\frac{(x-\alpha)^2}{2\sigma'^2}\right] dx$$

$$= 4\int_{\rho/\sigma'}^{\rho/\sigma} \frac{1}{\sqrt{2\pi}} e^{-u^2/2}\, du. \qquad (4\text{-}26)$$

Similarly, if $\sigma' < \sigma$,

$$I_1 = 4 \int_{\rho/\sigma}^{\rho/\sigma'} \frac{1}{\sqrt{2\pi}} e^{-u^2/2} du. \tag{4-27}$$

Combining Eqs. (4-26) and (4-27),

$$I_1 = 4 \left| \int_{\rho/\sigma}^{\rho/\sigma'} \frac{1}{\sqrt{2\pi}} e^{-u^2/2} du \right|, \tag{4-28}$$

where ρ is given by Eq. (4-23). From Eq. (4-23) we have

$$\frac{\rho}{\sigma} = \sigma' \left(\frac{2 \log(\sigma'/\sigma)}{\sigma'^2 - \sigma^2} \right)^{1/2}, \quad \frac{\rho}{\sigma'} = \sigma \left(\frac{2 \log(\sigma'/\sigma)}{\sigma'^2 - \sigma^2} \right)^{1/2}. \tag{4-29}$$

In terms of the error function we can write

$$I_1 = 2 \left| \operatorname{erf}\left(\frac{\rho}{\sigma' \sqrt{2}} \right) - \operatorname{erf}\left(\frac{\rho}{\sigma \sqrt{2}} \right) \right|. \tag{4-30}$$

Owing to the complexity of the exact forms (4-22) and (4-28), the invariant I_1 will not be of much use in significance tests. The exact form of I_1 when α and σ both vary will be extremely complicated.

4.6.2 Parameters of location and scale

For a distribution depending upon a location parameter α and a scale parameter σ, we have

$$p = \frac{1}{\sigma} f\left(\frac{x-\alpha}{\sigma}\right). \tag{4-31}$$

a. Suppose σ is known but α is unknown. For small variations of α the differential form of I_1 is

$$I_1 = d\alpha \int_{-\infty}^{\infty} \left| \frac{\partial p}{\partial \alpha} \right| dx$$

$$= \frac{d\alpha}{\sigma^2} \int_{-\infty}^{\infty} \left| f'\left(\frac{x-\alpha}{\sigma}\right) \right| dx$$

$$= \frac{d\alpha}{\sigma} \int_{-\infty}^{\infty} |f'(u)| \, du.$$

Thus $I_1 \propto d\alpha$, (4-32)

so that I_1 leads to the usual prior-probability form for a location parameter.

b. Suppose α is known but σ is unknown. For variations of σ the differential form of I_1 is

$$I_1 = d\sigma \int_{-\infty}^{\infty} \left|\frac{\partial p}{\partial \sigma}\right| dx$$

$$= \frac{d\sigma}{\sigma^2} \int_{-\infty}^{\infty} \left| f\left(\frac{x-\alpha}{\sigma}\right) + \frac{x-\alpha}{\sigma} f'\left(\frac{x-\alpha}{\sigma}\right) \right| dx$$

$$= \frac{d\sigma}{\sigma} \int_{-\infty}^{\infty} \left| f(u) + u\, f'(u) \right| du.$$

Hence $I_1 \propto \frac{d\sigma}{\sigma}$, (4-33)

which is the appropriate prior-probability form for a scale parameter.

4.6.3 The case when the parameter is a chance

Here we can regard x as a discrete variate assuming the values 0 and 1 with corresponding probabilities $1-\alpha$ and α.

$$I_1 = |(1 - \alpha') - (1 - \alpha)| + |\alpha' - \alpha|$$

$$= 2|\alpha' - \alpha| \qquad (4\text{-}34)$$

The differential form of I_1 is

$$I_1 = 2\, d\alpha \propto d\alpha, \qquad (4\text{-}35)$$

so that I_1 leads to the uniform prior-probability distribution of α.

4.6.4 Simple contingency

In this case I_1 leads to the prior-probability form of the parameter γ on the hypothesis q' in the significance test for γ (See Jeffreys, 1948, p. 232).

On the hypothesis q we have the set of chances

$$\begin{bmatrix} \alpha\beta & \alpha\beta' \\ \alpha'\beta & \alpha'\beta' \end{bmatrix} \qquad (4\text{-}36)$$

where $\alpha + \alpha' = 1 = \beta + \beta'$. On q' the set of chances is

$$\begin{bmatrix} \alpha\beta + \gamma & \alpha\beta' - \gamma \\ \alpha'\beta - \gamma & \alpha'\beta' + \gamma \end{bmatrix}. \tag{4-37}$$

Hence, from Eqs. (4-36) and (4-37),

$$I_1 = |\gamma| + |\gamma| + |\gamma| + |\gamma| = 4|\gamma|. \tag{4-38}$$

$$dI_1 \propto d\gamma. \tag{4-39}$$

Take $P(d\gamma | q', \alpha, \beta, H) \propto dI_1 \propto d\gamma$.

And since γ lies in the range $(-\alpha\beta, \alpha\beta')$, we have

$$P(d\gamma | q', \alpha, \beta, H) = \frac{d\gamma}{\alpha}, \tag{4-40}$$

where α is the least of α, α', β, or β'.

4.7 The Invariant I_m (m odd)

B. Raja Rao (1959a, 1959b) has investigated, under my supervision, the properties of I_m when m is odd. Though in this case it is not possible to obtain I_m explicitly in terms of parameters for all distributions admitting sufficient statistics, he has obtained an interesting convenient expression for it in the form

$$\begin{aligned} I_m = {} & \sum_{r=0}^{m} (-1)^r \binom{m}{r} \exp\left[\frac{m-r}{m} B' + \frac{r}{m} B - b\left(\frac{m-r}{m} u'_k + \frac{r}{m} u_k\right)\right] \\ & - 2 \sum_{r=0}^{m} (-1)^r \binom{m}{r} \exp\left[\frac{m-r}{m} B' + \frac{r}{m} B - b^*\left(\frac{m-r}{m} u'_k + \frac{r}{m} u_k\right)\right], \end{aligned} \tag{4-41}$$

where the function $b^*(u_k)$ is defined by

$$\int_{R^*} \exp\left[\sum_{k=1}^{p} u_k(\alpha_j) v_k(x) + A(x)\right] dx = \exp[-b^*(u_k)], \tag{4-42}$$

and R^* is the set of values of x satisfying the inequality

$$\sum_{k=1}^{p} (u'_k - u) v_k(x) \le B - B'. \tag{4-43}$$

Rao has verified that for many particular distributions admitting sufficient statistics b* can be evaluated. Note that Eq. (4-42) is analogous to Eq. (4-4). It is interesting to note that the first series in Eq. (4-41) is just the exact form of I_m (m even) obtained in Eq. (4-18). We shall now give some results obtained by Rao.

For the normal distribution with mean α and standard deviation σ, when both α and σ vary, the exact form of I_m (m odd) is

$$I_m = \pm \sum_{r=0}^{m} \frac{(-1)^r \binom{m}{r} m^2}{\sigma\sigma' r(m-r)} \delta_r$$

$$\times \exp\left[\frac{-r(m-r)(\alpha'-\alpha)^2 \delta_r^2}{2m^2\sigma^2\sigma'^2}\right]\left[2\int_{u_1}^{u_2} \frac{1}{\sqrt{2\pi}} e^{-u^2/2}\, du - 1\right], \qquad (4\text{-}44)$$

where the + or - sign is to be taken according to $\sigma > \sigma'$ or $\sigma < \sigma'$, and where

$$\delta_r^2 = \frac{m\sigma^2\sigma'^2}{(m-r)\sigma^2 + r\sigma'^2}$$

$$u_1 = \frac{(\alpha'-\alpha)\delta_r - K}{(\sigma^2 - \sigma'^2)\delta_r}$$

$$u_2 = \frac{(\alpha'-\alpha)\delta_r + K}{(\sigma^2 - \sigma'^2)\delta_r}$$

$$K = +[\rho^2(\sigma^2 - \sigma'^2)^2 + (\alpha'-\alpha)^2\, \sigma^2\sigma'^2]^{1/2}$$

$$\rho^2 = \frac{2\sigma^2\sigma'^2 \log(\sigma'/\sigma)}{\sigma'^2 - \sigma^2}.$$

In both the cases $\sigma > \sigma'$ and $\sigma < \sigma'$, I_m is obtained by taking the absolute value of the expression in Eq. (4-44).

When the parameters α and σ vary separately, the differential forms of $(I_m)^{1/m}$ satisfy

$$(I_m)^{1/m} \propto d\alpha \qquad (4\text{-}45)$$

and $$(I_m)^{1/m} \propto \frac{d\sigma}{\sigma}, \qquad (4\text{-}46)$$

which are the usual prior-probability forms. These results hold good even when m is even. But the difficulty arises when both the parameters vary simultaneously.

Rao has also obtained the exact forms of I_m (m odd) for the Type III distribution considered in Sec. 4.31, for the binomial and Poisson distributions. He has shown, as a particular case, that the differential forms of I_1 do not give the usual prior-probability forms for the binomial and the Poisson distributions.

5. Invariants of Distributions Admitting Sufficient Statistics: 1. One Parameter

5.1 Introduction

In this and the next three sections a theory of new invariants of distributions admitting sufficient statistics (or more generally an augmented sufficient set of statistics referred to in Sec. 2.4) is developed, and is applied in stating the prior probability of parameters in estimation problems.

As mentioned in Sec. 3.4, the invariance rule given by Jeffreys, though consistent, unfortunately does not lead to the appropriate prior-probability forms in many cases, and the necessity of further research has been stressed by Jeffreys himself. The new invariants that we are going to obtain enable us to arrive at all the appropriate prior-probability forms used by Jeffreys in those cases where sufficient statistics (or an augmented sufficient set of statistics) exist. For a single parameter it is possible to incorporate the results into an invariance rule. In the case of several parameters, though it has not been possible to give a single invariance rule which will be applicable to a large number of distributions, all the appropriate prior-probability forms will be expressed in terms of invariants. A satisfactory feature of our new invariants is that in all cases they lead to the uniform prior-probability distribution for a parameter which may have any value in a finite range or from $-\infty$ to $+\infty$, and for a location parameter in particular. They also lead to the prior-probability form $d\alpha/\alpha$ for a parameter α ranging from 0 to ∞, which includes the scale parameter as a special case (see Sec. 1.7).

There is one interesting point about consistency which needs mention here. There are some cases in which we can set up a model to explain a probability law. For instance, the parameter of the binomial distribution is the chance of the occurrence of an event in a single trial. The correlation coefficient also admits an interpretation in terms of a chance (Jeffreys, 1948, Sec. 2.5). In such cases it is necessary, for consistency, that the prior probability of the parameter in the law should be the same as the prior probability of the parameter in the model. An invariance rule therefore should lead to the same prior probability in both cases. The invariance rule given in Sec. 8 leads to the uniform prior probability in all cases when the parameter is a chance, or is the parameter of the binomial distribution, or is the correlation coefficient.

It may be emphasized here that it is not logically necessary to produce a single invariance rule which will be satisfactorily applicable to all distributions. In fact there seems to be little hope of ever discovering such a rule of universal application, just as it is unlikely to discover a single scientific law to explain satisfactorily all physical phenomena. *What is really important and essential, however, is that the prior-probability form that we wish to use should be expressed as an invariant, so that the requirement of consistency will be satisfied.* Hence there is nothing wrong even if we give different invariants for different distributions or combine the different invariants into a set of invariance rules instead of a single invariance rule. However, for the sake of simplicity it is advisable, as has been stressed by Jeffreys, to have as few rules as possible.

The general scheme of our work is to investigate all available invariants of distributions admitting sufficient statistics, and apply some of them to yield the satisfactory prior-probability forms considered by Jeffreys. Some of the remaining invariants will also be useful if the previous knowledge in those cases is such as to sanction the use of the prior-probability forms to which they lead.

The theory of new invariants, though developed primarily with a view of applications to prior probability, will be found interesting in itself, regarded as a study of some purely mathematical properties of the class of distributions admitting sufficient statistics. The situation is analogous to development of modern abstract geometry without consideration of applications to the real world. As mentioned in Sec. 3.2, some adherents of the frequency theory of probability have occasionally used invariants in their work in statistics, and it is not unlikely that the new invariants may find applications in their work also.

In this section we shall confine ourselves to distributions with a single parameter.

5.2 Standard Form

Let our distributions be of the type

$$f(x, \alpha) = \exp[u(\alpha)\ v(x) + A(x) + B(\alpha)]. \qquad (5\text{-}1)$$

The form (5-1) will be called the "standard form" of distributions under consideration. The functions, u, v, A, and B are, however, not unique when a distribution is written in the standard form (5-1). In general we can associate any *additive* and *multiplicative* constants with u and v. Thus Eq. (5-1) can also be written as

$$\begin{aligned} f(x, \alpha) &= \exp\ (\lambda_1 u + \mu_1)(\lambda_2 v + \mu_2) + (A - \mu_1\lambda_2 v) \\ &\quad + (B - \lambda_1\mu_2 u - \mu_1\mu_2) \end{aligned} \qquad (5\text{-}2)$$

where λ_1, μ_1, λ_2, and μ_2 are arbitrary constants subject only to the restriction $\lambda_1\lambda_2 = 1$. We shall be interested only in the function $u(\alpha)$, which will form the basis of our invariance theory. $u(\alpha)$ can be recognized as the coefficient of $v(x)$. In general $u(\alpha)$ may be replaced by $\lambda_1\ u(\alpha) + \mu_1$, λ_1 and μ_1 being any constants, without affecting the standard form (5-1). It follows that if a distribution is written in the standard form in any two ways such as Eq. (5-1) and

$$f(x, \alpha) = \exp[u_1(\alpha)\ v_1(x) + A_1(x) + B_1(\alpha)], \tag{5-3}$$

then $u_1 = \lambda u + \mu$, where λ and μ are constants.

5.2.1 Standard ranges

The three ranges $(-\infty, \infty)$, $(0, \infty)$, and $(-1, 1)$ will be regarded as "standard ranges." When $u(\alpha)$ has one of the three standard ranges, it will be said to have a standard range. A distribution will be said to be written in the "standard form with standard range" if it is written in the standard form in such a way that $u(\alpha)$ has a standard range.

We can always write a distribution in the standard form with standard range. For in choosing $u(\alpha)$ we have an additive and a multiplicative constant at our disposal. If the range of $u(\alpha)$ is unlimited in both directions, it is already a standard range $(-\infty, \infty)$. If it is unlimited in one direction only: if it is (k, ∞), the range of $u_1 = u - k$ is $(0, \infty)$; and if it is $(-\infty, k)$, the range of $u_1 = \lambda u + \mu$ is $(0, \infty)$, where $\lambda = -1$ and $\mu = k$. If it is limited in both directions, say (k_1, k_2), then $u_1 = \lambda u + \mu$ has the range $(-1, 1)$, where $\lambda = 2/(k_2 - k_1)$ and $\mu = -(k_2 + k_1)/(k_2 - k_1)$.

It may be observed that even if a distribution is written in the standard form with standard range, $u(\alpha)$ is not unique. But then we have the following lemma.

LEMMA 1

If a distribution is written in the standard form with standard range in any two ways giving $u(\alpha)$ and $u_1(\alpha)$ respectively, then

1. $u_1 = \lambda u + \mu$ if the standard range is $(-\infty, \infty)$,
2. $u_1 = \lambda u$ if the standard range is $(0, \infty)$,
3. $u_1 = \pm u$ if the standard range is $(-1, 1)$,

λ and μ being constants.

Proof. Item 1 needs no proof. To prove item 2, we have in general $u_1 = \lambda u + \mu$. Since the range of u is $(0, \infty)$, the range of $\lambda u + \mu$ is (μ, ∞). But u_1 has the same standard range $(0, \infty)$. Hence $\mu = 0$.

(Incidentally it may be observed that the character of a range is unaltered by operations with additive and multiplicative constants. Thus the range of $u_1 = \lambda u + \mu$ is unlimited in both directions or one direction or limited in both directions, according as the range of u is unlimited in both directions or one direction or limited in both directions. It is this fact which enables us to arrange for the same standard range for both u and u_1.)

To prove item 3. Let $u_1 = \lambda u + \mu$. If u has the range (-1, 1), the range of $\lambda u + \mu$ will be either $(-\lambda + \mu, \lambda + \mu)$ or $(\lambda + \mu, -\lambda + \mu)$. Hence either $-\lambda + \mu = -1$ and $\lambda + \mu = 1$ or else $-\lambda + \mu = 1$ and $\lambda + \mu = -1$, giving $\mu = 0$ and $\lambda = \pm 1$.

5.3 Transformations

Consider any transformation of the parameter and any transformation of the variate:

$$\alpha = \phi(\alpha'), \quad x = \psi(x').$$

The p.d.f. of x' is given by

$$f'(x', \alpha') = \exp\{u[\phi(\alpha')]\, v[\psi(x')] + A[\psi(x')] + B[\phi(\alpha')]\} \left|\frac{d\psi}{dx'}\right| .$$

Or writing in the standard form,

$$f'(x', \alpha') = \exp\{u[\phi(\alpha')]\, v[\psi(x')] + A[\psi(x')] + \log\left|\frac{d\psi}{dx'}\right| + B[\phi(\alpha')]\}. \tag{5-4}$$

We note that $f(x, \alpha)$ and $f'(x', \alpha')$ have the same form. *Hence the form of our distributions is invariant for all transformations of the parameter and the variate.*

As before there is choice of additive and multiplicative constants in writing $f'(x', \alpha')$ in the standard form. If $f'(x', \alpha')$ is written in the standard form in any other way as

$$f'(x', \alpha') = \exp[u'(\alpha')\, v'(x') + A'(x') + B'(\alpha')], \tag{5-5}$$

then we shall have, as in Sec. 5.2,

$$u'(\alpha') = \lambda\, u[\phi(\alpha')] + \mu = \lambda\, u(\alpha) + \mu, \tag{5-6}$$

where λ and μ are constants. If, after the transformation, $f'(x', \alpha')$ is also written in the standard form with standard range, one or both of the constants λ and μ can be determined in some cases. This leads to the following lemma.

LEMMA 2

If both $f(x, \alpha)$ and $f'(x', \alpha')$ are written in the standard form with standard range and $u(\alpha)$ and $u'(\alpha')$ are corresponding functions in the two forms, then for corresponding values of α and α' we have

$u'(\alpha') = \lambda\, u(\alpha) + \mu$ if the standard range is $(-\infty, \infty)$,

$u'(\alpha') = \lambda\, u(\alpha)$ if the standard range is $0, \infty)$,

$u'(\alpha') = \pm\, u(\alpha)$ if the standard range is $(-1, 1)$.

This may be written as $|u'(\alpha')| = |u(\alpha)|$.

The proof is exactly the same as that for Lemma 1 of Sec. 5.2.1.

5.4 Types of Invariants

In this section we shall frame some terminology which will be useful in our work. Suppose that a certain rule enables us to identify a function $u(\alpha)$ uniquely, except possibly for a multiplicative or additive constant. Suppose that there is some transformation carrying α to α'. If the same rule enables us to identify after transformation a function $u'(\alpha')$ uniquely, except again for a multiplicative or additive constant, and if for corresponding values of α and α' we have the following relations:

1. If $u'(\alpha') = \lambda\, u(\alpha) + \mu$, where λ and μ are constants, then $u(\alpha)$ will be called an invariant of Type III with regard to that transformation.

2. If $u'(\alpha') = \lambda\, u(\alpha)$, we shall call $u(\alpha)$ an invariant of Type II.

3. If $u'(\alpha') = u(\alpha) + \mu$, then $u(\alpha)$ will be called an invariant of Type III.

4. If $u'(\alpha') = u(\alpha)$, then $u(\alpha)$ will be said to be an invariant of Type I.

An invariant of Type I may be compared to what is called an "absolute invariant" in the theory of algebraic invariants, and that of Type II to a "relative invariant."

If u is an invariant of Type I, $\phi(u)$ is also an invariant of Type I where ϕ is any arbitrary function. $d\phi(u)$ (d denoting the differential) is also an invariant and may be called a differential invariant of Type I.

If u is an invariant of Type II, we have

$$u' = \lambda u, \quad du' = \lambda\, du$$

so that du is a differential invariant of Type II. Again $du'/u' = du/u$. Hence du/u is a differential invariant of Type I. Moreover, $u'^{\ell+1} = \lambda^{\ell+1} u^{\ell+1}$ (ℓ being any constant not equal to -1) and $u'^{\ell}\, du' = \lambda^{\ell+1} u^{\ell}\, du$, so that $u^{\ell}\, du$ is a differential invariant of Type II. This is apparently the most general differential invariant that we can have in this case.

For an invariant of Type II[1],

$$u' = u + \mu, \quad du' = du,$$

so that du is a differential invariant of Type I. This is the only differential invariant that we can have in this case.

If u is an invariant of Type III,

$$u' = \lambda u + \mu, \quad du' = \lambda\, du.$$

Hence du is a differential invariant of Type II. This again is the only differential invariant in this case.

For applications to the prior probability we need differential invariants. In stating the prior probability of a parameter any unspecified multiplicative constant is immaterial, since it can be left to the observations to be decided. Hence differential invariants of both Types I and II are useful in stating the prior probability of parameters. Though from the mathematical point of view there is a difference between these two types, there is no real difference from the standpoint of applications to the prior probability. Consequently in the discussion of the applications of differential invariants to the prior probability the Type of an invariant will be often omitted.

Though all the four types of invariants yield differential invariants, it is important to observe that an invariant of Type I can lead to any desired prior-probability form. The most general prior-probability form to which an invariant of Type II leads is $u^{\ell}\,du$, ℓ being any constant. With invariants of Types II^1 and III the only available prior-probability form is du. There is again no real difference between invariants of Types II^1 and III while considering applications to the prior probability.

5.5 Invariants

With the terminology of the preceding section the results of Lemma 2 in Sec. 5.3 can be restated as follows:

When a given distribution is *always* written in the standard form with standard range, for all transformations of the parameter and the variate.

1. u is an invariant of Type III if its standard range is $(-\infty, \infty)$, and du is a differential invariant of Type II.

2. u is an invariant of Type II if its standard range is $(0, \infty)$. du/u is a differential invariant of Type I and $u^{\ell}\,du$ (ℓ being any constant not equal to -1) is a differential invariant of Type II.

3. $|u|$ is an invariant of Type I if the standard range of u is $(-1, 1)$. $\phi(|u|)$ is also an invariant of Type I where ϕ is any arbitrary function, and $d\phi(|u|)$ is a differential invariant of Type I.

5.6 Applications to the Prior Probability

At first we shall state an invariance rule for the prior probability of a parameter and test it for some well-known distributions. Cases where it is desired to have other prior-probability forms than those given by the rule will be considered later.

Invariance rule. Write the given distribution in the standard form with standard range. Take $P(du|H) \propto du$ if the standard range of u is $(-\infty, \infty)$ or $(-1, 1)$, and $P(du|H) \propto du/u$ if the standard range of u is $(0, \infty)$.

If this rule does not lead to satisfactory prior-probability forms in certain cases, there is complete choice for its modification if the second range of u is (-1, 1), and a partial choice of the form $u^{\ell}\,du$ if the standard range is $(0, \infty)$.

5.6.1 The normal distribution

$$y = \frac{1}{\sigma\sqrt{2\pi}} \exp\left[- \frac{(x-\alpha)^2}{2\sigma^2} \right].$$

a. If σ is known,

$$f(x, \alpha) = \exp\left(\frac{\alpha x}{\sigma^2} - \frac{x^2}{2\sigma^2} - \frac{\alpha^2}{2\sigma^2} - \log \sigma\sqrt{2\pi} \right).$$

Here $u(\alpha) = \alpha$ has the standard range $(-\infty, \infty)$. By the rule,

$$P(du|H) \propto du = d\alpha.$$

Hence the rule leads to

$$P(d\alpha|H) \propto d\alpha,$$

which is satisfactory.

b. If α is known but σ unknown,

$$f(x, \sigma) = \exp\left[- \frac{(x-\alpha)^2}{2\sigma^2} - \log \sigma\sqrt{2\pi} \right].$$

Here $u(\sigma) = 1/\sigma^2$ has the standard range $(0, \infty)$. Hence, by the rule,

$$P(du|H) \propto \frac{du}{u} = -2\,\frac{d\sigma}{\sigma}.$$

Thus the rule leads to

$$P(d\sigma|H) \propto \frac{d\sigma}{\sigma},$$

the usual form.

5.6.2 The Poisson distribution

$$f(x, \alpha) = \frac{e^{-\alpha}\alpha^x}{x!} = \exp(x \log \alpha - \log x! - \alpha).$$

Here $u(\alpha) = \log \alpha$ has the standard range $(-\infty, \infty)$, since the range of

α is $(0, \infty)$. $du = d\alpha/\alpha$. Hence the rule leads to the desired prior-probability form $d\alpha/\alpha$.

5.6.3 The binomial distribution

$$f(x, \alpha) = \binom{n}{x}\alpha^x(1-\alpha)^{n-x}$$

$$= \binom{n}{x}\left(\frac{\alpha}{1-\alpha}\right)^x(1-\alpha)^n$$

$$= \exp\left[x \log\left(\frac{\alpha}{1-\alpha}\right) + \log\binom{n}{x} + n \log(1-\alpha)\right].$$

$u(\alpha) = \log[\alpha/(1-\alpha)]$ ranges from $-\infty$ to ∞, since α ranges from 0 to 1. $du = d\alpha/\alpha(1-\alpha)$. Hence the rule leads to the prior-probability form $d\alpha/\alpha(1-\alpha)$, which has been advocated by Haldane (1932). This form gives infinite density at the limits and is satisfactory if there is reason to believe that the population sampled is homogeneous. For ordinary purposes, however, the uniform assessment of the prior probability is satisfactory. Though this rule does not lead to the form $d\alpha$, we shall find later an invariant for that form.

5.6.4 When the parameter α is a chance

Here we can conveniently regard α as the parameter of the distribution for which $f(x, \alpha) = \alpha^{1-x}(1-\alpha)^x$, where x takes only the values 0 and 1. Then

$$f(x, \alpha) = \alpha\left(\frac{1-\alpha}{\alpha}\right)^x$$

$$= \exp\left[x \log\left(\frac{1-\alpha}{\alpha}\right) + \log \alpha\right].$$

$u(\alpha) = \log[(1-\alpha)/\alpha]$ ranges from $-\infty$ to ∞. $du = -\, d\alpha/\alpha(1-\alpha)$, so that the rule gives again Haldane's form.

5.6.5 Pearson's Type III distribution

$$y = \frac{a^p e^{-ax} x^{p-1}}{\Gamma(p)} \qquad (0 \le x < \infty,\ a > 0,\ p > 0).$$

a. If a is known,

$$f(x, p) = \exp[(p-1) \log x - ax + p \log a - \log \Gamma(p)].$$

Here $u(p) = p-1$ has the range $(-1, \infty)$, which is not a standard range. Writing $f(x, p)$ so that $u(p)$ has a standard range, we have

$$f(x, p) = \exp[p \log x - \log x - ax + p \log a - \log \Gamma(p)].$$

Now $u(p) = p$ has the standard range $(0, \infty)$. $du/u = dp/p$, so that the rule leads to the prior-probability form dp/p, which is the appropriate form following Jeffreys.

b. If p is known but a is unknown,

$$f(x, a) = \exp[- ax + (p-1) \log x + p \log a - \log \Gamma(p)].$$

$u(a) = a$ has the standard range $(0, \infty)$, and we have the form da/a.

5.6.6 The beta distribution

$$y = \frac{x^{\ell-1}(1-x)^{m-1}}{B(\ell, m)} \qquad (0 \leq x \leq 1;\ \ell > 0,\ m > 0).$$

If m is known, the rule gives the form $d\ell/\ell$. If ℓ is known, it gives the form dm/m for m.

5.6.7 Pearson's Type VIII distribution

$$y = k\left(1 + \frac{x}{a}\right)^{-m} \qquad (-a \leq x \leq 0;\ -\infty < m < 1),$$

and k is a function of a and m. If a is known,

$$f(x, m) = \exp\left[(1-m) \log\left(1 + \frac{x}{a}\right) - \log\left(1 + \frac{x}{a}\right) + \log k\right].$$

$u(m) = 1-m$ has the standard range $(0, \infty)$. $du/u = - dm/(1-m)$, giving the appropriate form $dm/(1-m)$.

5.6.8 Pearson's Type XII distribution

$$y = k\left[\frac{1 + (x/a_1)}{1 - (x/a_2)}\right]^m \qquad (-a_1 \leq x \leq a_2;\ |m| < 1),$$

and k is a function of a_1, a_2, and m. If a_1 and a_2 are known,

$$f(x, m) = \exp\left\{m \log\left[\frac{1 + (x/a_1)}{1 - (x/a_2)}\right] + \log k\right\}.$$

Here $u(m) = m$ has the standard range $(-1, 1)$. $du = dm$, giving

$P(dm|H) = \frac{1}{2}\,dm$. This distribution is interesting in that we can find an invariant for any desired prior-probability form. $|u(m)|$ is an invariant of Type I, since it has the range (-1, 1). $\phi(|u|)$ and $d\,\phi(|u|)$ are also invariants of Type I, where ϕ is any arbitrary function. The rule gives satisfactory results for other Pearson distributions also, if we conaider the appropriate parameter for which a sufficient statistic exists, the remaining parameters being supposed known.

5.7 Combination of Invariants

The invariants obtained by us can be combined with the general invariant J so as to yield new invariants. For a single parameter, $J = g_{11}\,d\alpha^2$, and this is a differential invariant of Type I, true for any distribution. By Eq. (4-14) J when expressed in terms of u is

$$J = -\frac{\partial^2 B}{\partial u^2}\,du^2 = G_{11}\,du^2. \quad \left(G_{11} = -\frac{\partial^2 B}{\partial u^2}\right).$$

Since $J = G_{11}\,du^2$ is a differential invariant of Type I and du is itself a differential invariant of Type II when the standard range of u is $(-\infty, \infty)$, it follows by division that $J/du = G_{11}\,du$ is a differential invariant of Type II. Similarly, if the standard range of u is $(0, \infty)$, du/u is a differential invariant of Type I and $Ju/du = G_{11}u\,du$ is also a differential invariant of Type I. It is not necessary to consider the case when the standard range of u is (-1, 1), since there is already complete choice in that case.

For the binomial distribution, $J = d\alpha^2/\alpha(1-\alpha)$ and $du = d\alpha/\alpha(1-\alpha)$, so that $J/du = d\alpha$. Thus the invariant J/du gives uniform prior probability for α. The same result holds good when the parameter α is a chance.

For the normal distribution the new invariants combined with J lead to the usual forms. But for the Poisson distribution they give the form $d\alpha$ instead of the usual form $d\alpha/\alpha$.

6. Invariants of Distributions Admitting Sufficient Statistics: 2. Several Parameters

6.1 Introduction

In this section we shall obtain invariants of distributions with several parameters and consider their applications to the prior probability. More generally, we shall consider distributions admitting an augmented sufficient set of statistics. Results that we shall obtain hold good for multivariate distributions also where we have to replace x simply by x_q.

6.2 Standard Form

Let our distributions be of the type

$$f(x, \alpha_j) = \exp\left[\sum_{k=1}^{\nu} u_k(\alpha_j)\ v_k(x) + A(x) + B(\alpha_j)\right], \tag{6-1}$$

where ν may be greater than or equal to p, the number of parameters α_j.

The form (6-1) will be called the "standard form" of distributions under consideration, provided ν is the least integer for which $f(x, \alpha_j)$ can be written as in Eq. (6-1), and further none of the u's contains a term (other than a numerical constant) which is a multiple of a term contained in any other u, and none of the v's contains a term (other than a numerical constant) which is a multiple of a term contained in any other v; e.g., the normal distribution

$$f(x, \alpha, \sigma) = \frac{1}{\sigma\sqrt{2\pi}} \exp\left[-\frac{(x-\alpha)^2}{2\sigma^2}\right],$$

can be written in two forms as

$$f(x, \alpha, \sigma) = \exp\left(-\frac{x^2}{2\sigma^2} + \frac{\alpha x}{\sigma^2} - \frac{\alpha^2}{2\sigma^2} - \log \sigma\sqrt{2\pi}\right), \tag{6-2}$$

and

$$f(x, \alpha, \sigma) = \exp\left[\left(-\frac{1}{2\sigma^2} - \frac{\alpha}{\sigma^2}\right)x^2 + \frac{\alpha}{\sigma^2}(x^2 + x) - \frac{\alpha^2}{2\sigma^2} - \log \sigma\sqrt{2\pi}\right]. \tag{6-3}$$

The form (6-2) is a standard form. But in the form (6-3), taking

$$u_1 = -\frac{1}{2\sigma^2} - \frac{\alpha}{\sigma^2}, \qquad u_2 = \frac{\alpha}{\sigma^2},$$

$$v_1 = x^2, \qquad v_2 = x^2 + x,$$

u_2 contains the term α/σ^2, which is a multiple of a term in u_1; the term x^2 in v_1 is again a term in v_2. Consequently Eq. (6-3) is not a standard form.

LEMMA 1

Whatever be the mode of writing $f(x, \alpha_j)$ in the standard form, the functions $u_k(\alpha_j)\ v_k(x)$ will be unique provided the u_k and the v_k do not contain any explicit additive constants.

Proof. Let $f(x, \alpha_j)$ be written in the standard form in any two ways such as Eq. (6-1) and

$$f(x, \alpha_j) = \exp\left[\sum_{k=1}^{\nu} u_k'(\alpha_j)\ v_k'(x) + A'(x) + B'(\alpha_j)\right]. \tag{6-4}$$

ν, being the least integer for a standard form, will be the same for both Eqs. (6-1) and (6-4). Then for all x and α_j

$$\sum_{k=1}^{\nu} u_k(\alpha_j)\ v_k(x) + A(x) + B(\alpha_j)$$

$$\equiv \sum_{k=1}^{\nu} u_k'(\alpha_j)\ v_k'(x) + A'(x) + B'(\alpha_j). \tag{6-5}$$

Consider any one of the $u_k(\alpha_j)\ v_k(x)$, say $u_1(\alpha_j)\ v_1(x)$. Let

$$u_1(\alpha_j) = \sum_{\ell} a_\ell(\alpha_j) \quad \text{and} \quad v_1(x) = \sum_m b_m(x).$$

Then $u_1(\alpha_j)\ v_1(x) = \left[\sum_{\ell} a_\ell(\alpha_j)\right]\left[\sum_m b_m(x)\right]$ (6-6)

Then term $a_1(\alpha_j)\ b_1(x)$ of $u_1(\alpha_j)\ v_1(x)$ will be contained in one of the $u_k'(\alpha_j)\ v_k'(x)$, say $u_1'(\alpha_j)\ v_1'(x)$. Then $a_1(\alpha_j)$ will be contained in $u_1'(\alpha_j)$ and $b_1(x)$ will be contained in $v_1'(x)$. Any other term of $u_1(\alpha_j)\ v_1(x)$ such as $a_1(\alpha_j)\ b_m(x)$ must also be contained in the same

$u'_1(\alpha_j)\ v'_1(x)$ for every m. If not, suppose $a_1(\alpha_j)\ b_m(x)$ is contained in some other $u'_k(\alpha_j)\ v'_k(x)$, say $u'_2(\alpha_j)\ v'_2(x)$. Then $a_1(\alpha_j)$ will occur in $u'_2(\alpha_j)$. Hence the term $a_1(\alpha_j)$ occurs both in $u'_1(\alpha_j)$ and $u'_2(\alpha_j)$, which is contrary to our definition of a standard form that none of the $u'_k(\alpha_j)$ contains a term which is a multiple of a term in some other $u'_k(\alpha_j)$.

Every term such as $a_1(\alpha_j)\ b_m(x)$ therefore occurs in $u'_1(\alpha_j)\ v'_1(x)$. Further, $b_m(x)$ occurs in $v'_1(x)$, and similar argument shows that all terms such as $a_\ell(\alpha_j)\ b_m(x)$ must also occur in $u'_1(\alpha_j)\ v'_1(x)$. All terms of $u_1(\alpha_j)\ v_1(x)$ are therefore contained in $u'_1(\alpha_j)\ v'_1(x)$. Arguing round the other way, all terms of $u'_1(\alpha_j)\ v'_1(x)$ must also occur in $u_1(\alpha_j)\ v_1(x)$.

Hence $u_1(\alpha_j)\ v_1(x) \equiv u'_1(\alpha_j)\ v'_1(x)$ so that $u_1(\alpha_j)\ v_1(x)$ is unique, and similarly each $u_k(\alpha_j)\ v_k(x)$ is unique.

There is, however, apparently no means of separating $u_k(\alpha_j)$ from $v_k(x)$ uniquely, owing to the ambiguity of a multiplicative constant, since any multiplicative constant can be associated with $u_k(\alpha_j)$. It is also convenient to relax the restriction in Lemma 1 that the u_k and the v_k do not contain explicit additive constants. When this is done, we shall have Lemma 2.

LEMMA 2

Whatever be the mode of writing $f(x, \alpha_j)$ in the standard form, the u_k will be unique except for multiplicative and additive constants.

This follows from Lemma 1, since any u_k can now be replaced by $\lambda_1 u_k + \mu_1$ without affecting the standard form, λ_1 and μ_1 being any constants; e.g.,

$$(\lambda_1 u_k + \mu_1)v_k = \lambda_1 u_k v_k + \mu_1 v_k ,$$

and $\mu_1 v_k$ can be included in A(x), being a function of x only.

It also follows that if $u_k(\alpha_j)$ and $u'_k(\alpha_j)$ are corresponding functions, corresponding to any two ways of writing $f(x, \alpha_j)$ in the standard form,

$$u_k'(\alpha_j) = \lambda\, u_k(\alpha_j) + \mu,$$

where λ and μ are constants.

One or both of the constants λ and μ can be fixed in some cases by the device of standard ranges for the u_k. The same ranges as in Sec. 5.2.1 will be taken as the standard ranges for the u_k. If $f(x, \alpha_j)$ is written in the standard form such that each u_k has a standard range, it will be said to be written in the "standard form with standard ranges." The further theory is much the same as that corresponding to it in the preceding section.

6.3 Transformations

Consider any transformation of the α_j into the α_j' and any transformation of x into x', transforming $f(x, \alpha_j)$ into

$$f'(x', \alpha_j') = \exp\left[\sum_{k=1}^{\nu} u_k'(\alpha_j')\, v_k'(x') + A'(x') + B'(\alpha_j')\right].$$

If both $f(x, \alpha_j)$ and $f'(x', \alpha_j')$ are written in the standard form with standard ranges, we shall have for corresponding values of the α_j and the α_j'

$$u_k'(\alpha_j') = \lambda\, u_k(\alpha_j) + \mu \quad \text{if the standard range of } u_k \text{ is } (-\infty, \infty),$$

$$u_k'(\alpha_j') = \lambda\, u_k(\alpha_j) \quad \text{if the standard range of } u_k \text{ is } (0, \infty),$$

$$u_k'(\alpha_j') = \pm\, u_k(\alpha_j) \quad \text{if the standard range of } u_k \text{ is } (-1, 1),$$

λ and μ being constants. The proof is exactly the same as that for Lemma 2 in Sec. 5.3.

6.4 Invariants

The results of Sec. 6.3 can be restated as follows: When a given distribution is always written in the standard form with standard ranges, for all transformations of the parameters and the variate,

1. u_k is an invariant of Type III if its standard range is $(-\infty, \infty)$. du_k is a differential invariant of Type II.

2. u_k is an invariant of Type II if its standard range is $(0, \infty)$. du_k/u_k is a differential invariant of Type I and $u_k^{\ell}\, du_k$ is a differential invariant of Type II, ℓ being any constant not equal to -1.

3. $|u_k|$ is an invariant of Type I if the standard range of u_k is (-1, 1). $\phi(|u_k|)$ is also an invariant of Type I where ϕ is any arbitrary function, and $d\,\phi(|u_k|)$ is a differential invariant of Type I.

6.5 Applications to the Prior Probability

As in the preceding section we can use differential invariants (types I and II) in stating the prior probability of parameters in estimation problems. For the prior-probability differential of p parameters we require an invariant which is a product of p differentials du_k. Thus for a distribution with three parameters suppose u_1 has the range $(0, \infty)$ and both u_2 and u_3 have the range $(-\infty, \infty)$. Then $u_1^{\ell}\, du_1$, du_2, and du_3 are differential invariants. By multiplication, $u_1^{\ell}\, du_1\, du_2\, du_3$ is also a differential invariant and may be taken as proportional to the prior-probability differential of the joint distribution of u_1, u_2, and u_3. The prior-probability differential of the joint distribution of α_1, α_2, and α_3 will then be proportional to

$$u_1^{\ell}\left|\frac{\partial(u_1,\, u_2,\, u_3)}{\partial(\alpha_1,\, \alpha_2,\, \alpha_3)}\right|\, d\alpha_1\, d\alpha_2\, d\alpha_3.$$

In the case of several parameters we do not get a single invariance rule which will be applicable to a large number of distributions. However, we can use different invariants to lead to different prior-probability forms. Our procedure will be to express the appropriate prior-probability forms for different distributions in terms of the invariants obtained earlier. We shall come across some cases where the above invariants based on the u's are not adequate by themselves, but then they can be combined with the general invariant J so as to yield new invariants which will be found to be adequate for our purposes.

General Requirement. While expressing the prior-probability forms in terms of invariants there is one general requirement which deserves attention. The standard ranges of the u_k's are themselves invariants of all transformations. All the u_k can be classified into three groups: (1) those having the standard range $(-\infty, \infty)$ (2) those having the standard range $(0, \infty)$, and (3) those having the standard range $(-1, 1)$. But when a distribution is given in any form, there is apparently no further means of identifying the individual u_k belonging to a particular group. Hence the invariant that we wish to use for the prior probability should be a symmetric function of all the u_k's belonging to the same group.

The point in question can be illustrated by examples. Suppose in a certain distribution with four parameters both u_1 and u_2 have the range $(-\infty, \infty)$ and both u_3 and u_4 have the range $(0, \infty)$. Suppose further that the expression for the appropriate prior-probability form is

$$\frac{du_1\, du_2\, du_3\, du_4}{u_3^2 u_4^3}.$$

No doubt this expression is an invariant, but it is not a symmetric function of u_3 and u_4, which belong to the same group. As there is no means of discriminating between u_3 and u_4, the invariant expression in question cannot be used in stating the prior probability of the parameters. On the other hand, there is no objection against an invariant such as

$$u_3^\ell u_4^\ell\, du_1\, du_2\, du_3\, du_4.$$

The point becomes important for distributions admitting an augmented sufficient set of statistics where ν, the number of the u_k's, exceeds p, the number of the parameters. For the prior probability we need just p of the ν differentials du_k. In choosing these p differentials we must take either all or none belonging to the same group. Suppose there are three u_k's for a certain distribution with two parameters. Suppose u_1, u_2, and u_3 all have the same range $(0, \infty)$ and the invariant expression for the prior probability comes out to be $du_1\ du_2/u_1u_2$. Such an expression cannot be used, since there is no means of selecting u_1 and u_2 from the trio. On the other hand, if both u_1 and u_2 have the range $(0, \infty)$ and u_3 has the range $(-\infty, \infty)$, there is no objection against the invariant $du_1\ du_2/u_1u_2$.

6.6 Prior-Probability Forms in Terms of Invariants

We now proceed to express some prior-probability forms (used by Jeffreys) in terms of invariants.

6.6.1 The normal distribution with two parameters

$$f(x, \alpha, \sigma) = \frac{1}{\sigma\sqrt{2\pi}} \exp\left[- \frac{(x-\alpha)^2}{2\sigma^2}\right]$$

$$= \exp\left[- \frac{x^2}{2\sigma^2} + \frac{\alpha x}{\sigma^2} - \frac{\alpha^2}{2\sigma^2} - \log \sigma\sqrt{2\pi}\right].$$

Here $u_1 = 1/\sigma^2$ has the standard range $(0, \infty)$ and $u_2 = \alpha/\sigma^2$ has the standard range $(-\infty, \infty)$.

$$\frac{\partial(u_1, u_2)}{\partial(\alpha, \sigma)} = \frac{2}{\sigma^5}.$$

Hence

$$du_1\ du_2 = \frac{2}{\sigma^5}\, d\alpha\ d\sigma$$

or

$$\frac{du_1\ du_2}{u_1^2} = \frac{2\ d\alpha\ d\sigma}{\sigma} \propto \frac{d\alpha\ d\sigma}{\sigma}.$$

Now $du_1\ du_2/u_1^2$ is an invariant, and if we take

$$P(du_1\ du_2|H) \propto \frac{du_1\ du_2}{u_1^2},$$

we shall have

$$P(d\alpha\ d\sigma|H) \propto \frac{d\alpha d\sigma}{\sigma}.$$

Hence the invariant $du_1\ du_2/u_1^2$ leads to the prior-probability form $d\alpha\ d\sigma/\sigma$ for the normal distribution. It may be remarked here that u_1 and u_2 can be distinguished from each other, since they have different ranges.

6.6.2 Pearson's Type III distribution

$$f(x, a, p) = \frac{a^p e^{-ax} x^{p-1}}{\Gamma(p)} \qquad (0 \leq x < \infty;\ a > 0,\ p > 0)$$

$$= \exp[-ax + p \log x - \log x + p \log a - \log \Gamma(p)].$$

Both $u_1 = a$ and $u_2 = p$ have the standard range $(0, \infty)$

$$\frac{du_1\ du_2}{u_1 u_2} = \frac{da\ dp}{ap}.$$

Hence the invariant $du_1\ du_2/u_1u_2$ gives the prior-probability form $da\ dp/ap$.

6.6.3 The beta distribution

$$f(x, \ell, m) = \frac{x^{\ell-1}(1-x)^{m-1}}{B(\ell, m)} \qquad (0 \leq x \leq 1;\ \ell > 0,\ m > 0).$$

Here again the invariant $du_1\ du_2/u_1u_2$ leads to the form $d\ell\ dm/\ell m$.

6.6.4 The bivariate normal distribution with known means

$$f(x, y, \sigma_1, \sigma_2, \rho) = \frac{1}{2\pi\sigma_1\sigma_2(1-\rho^2)^{1/2}}$$

$$\exp\left[-\frac{1}{2(1-\rho^2)}\left(\frac{x^2}{\sigma_1^2} - \frac{2\rho xy}{\sigma_1\sigma_2} + \frac{y^2}{\sigma_2^2}\right)\right]$$

$$= \exp\left[-\frac{x^2}{2\sigma_1^2(1-\rho^2)} - \frac{y^2}{2\sigma_2^2(1-\rho^2)}\right.$$

$$+ \frac{\rho xy}{\sigma_1\sigma_2(1-\rho^2)} - \log 2\pi\sigma_1\sigma_2(1-\rho^2)^{1/2}\Bigg].$$

Here both

$$u_1 = \frac{1}{\sigma_1^2(1-\rho^2)} \quad \text{and} \quad u_2 = \frac{1}{\sigma_2^2(1-\rho^2)}$$

have the range $(0, \infty)$.

$$u_3 = \frac{\rho}{\sigma_1\sigma_2(1-\rho^2)}$$

has the range $(-\infty, \infty)$.

$$\frac{\partial(u_1, u_2, u_3)}{\partial(\sigma_1, \sigma_2, \rho)} \propto \frac{1}{\sigma_1^4\ \sigma_2^4(1-\rho^2)^3}.$$

Hence

$$du_1\ du_2\ du_3 \propto \frac{d\sigma_1\ d\sigma_2\ d\rho}{\sigma_1^4\ \sigma_2^4(1-\rho^2)^3},$$

and

$$\frac{du_1\ du_2\ du_3}{u_1^{3/2}u_2^{3/2}} \propto \frac{d\sigma_1\ d\sigma_2\ d\rho}{\sigma_1\sigma_2},$$

since $u_1u_2 = \dfrac{1}{\sigma_1^2\ \sigma_2^2(1-\rho^2)^2}$.

Further $du_1\ du_2\ du_3/u_1^{3/2}u_2^{3/2}$ is an invariant. Hence, if we take

$$P(du_1\ du_2\ du_3|H) \propto \frac{du_1\ du_2\ du_3}{u_1^{3/2}\ u_2^{3/2}},$$

we shall have

$$P(d\sigma_1\ d\sigma_2\ d\rho\,|H) \propto \frac{d\sigma_1\ d\sigma_2\ d\rho}{\sigma_1\sigma_2}.$$

Note that our invariant is a symmetric function of u_1 and u_2, which belong to the same group.

6.6.5 The most general bivariate normal distribution

$$f(x, y, \alpha_1, \alpha_2, \sigma_1, \sigma_2, \rho) = \frac{1}{2\pi\sigma_1\sigma_2(1-\rho^2)^{1/2}} \exp\left[-\frac{1}{2(1-\rho^2)} \left[\frac{(x-\alpha_1)^2}{\sigma_1^2} - \frac{2\rho(x-\alpha_1)(y-\alpha_2)}{\sigma_1\sigma_2} + \frac{(y-\alpha_2)^2}{\sigma_2^2} \right]\right]$$

$$= \exp\left[-\frac{1}{2} u_1 x^2 - \frac{1}{2} u_2 y^2 + u_3 xy + u_4 x + u_5 y + B(\alpha_1, \alpha_2, \sigma_1, \sigma_2, \rho)\right],$$

where $u_1 = \dfrac{1}{\sigma_1^2(1-\rho^2)}$ and $u_2 = \dfrac{1}{\sigma_2^2(1-\rho^2)}$

both have the range $(0, \infty)$; and

$$u_3 = \frac{\rho}{\sigma_1\sigma_2(1-\rho^2)},$$

$$u_4 = \frac{1}{1-\rho^2}\left(\frac{\alpha_1}{\sigma_1^2} - \frac{\rho\alpha_2}{\sigma_1\sigma_2}\right),$$

and $$u_5 = \frac{1}{1-\rho^2}\left(\frac{\alpha_2}{\sigma_2^2} - \frac{\rho\alpha_1}{\sigma_1\sigma_2}\right),$$

each have the range $(-\infty, \infty)$. We have

$$\frac{\partial(u_1, u_2, u_3, u_4, u_5)}{\partial(\alpha_1, \alpha_2, \sigma_1, \sigma_2, \rho)} \propto \frac{1}{\sigma_1^6 \sigma_2^6 (1-\rho^2)^4},$$

so that $$du_1\, du_2\, du_3\, du_4\, du_5 \propto \frac{d\alpha_1\, d\alpha_2\, d\sigma_1\, d\sigma_2\, d\rho}{\sigma_1^6 \sigma_2^6 (1-\rho^2)^4}. \quad (6\text{-}7)$$

The appropriate prior-probability form in the present case is $d\alpha_1\, d\alpha_2\, d\sigma_1\, d\sigma_2\, d\rho/\sigma_1\sigma_2$. But this cannot be expressed as an invariant in terms of the u's. For the most general invariant that the u's can give for the prior probability is $u_1^\ell u_2^\ell\, du_1\, du_2\, du_3\, du_4\, du_5$, where ℓ is any constant. In virtue of Eq. (6-7) and using

$$u_1 u_2 = \frac{1}{\sigma_1^2 \sigma_2^2 (1-\rho^2)^2},$$

$$u_1^{\ell} u_2^{\ell}\, du_1\, du_2\, du_3\, du_4\, du_5 \propto \frac{d\alpha_1\, d\alpha_2\, d\sigma_1\, d\sigma_2\, d\rho}{\sigma_1^{2\ell+6}\, \sigma_2^{2\ell+6} (1-\rho^2)^{2\ell+4}}. \tag{6-8}$$

But the right-hand side of Eq. (6-8) does not reduce to

$$\frac{d\alpha_1\, d\alpha_2\, d\sigma_1\, d\sigma_2\, d\rho}{\sigma_1 \sigma_2}$$

for any value of ℓ. However, by combining the invariants of the u system with the general invariant J we can obtain an invariant for the prior-probability form in question. Taking $\ell = -3$ in Eq. (6-8) we have

$$\frac{du_1\, du_2\, du_3\, du_4\, du_5}{u_1^3 u_2^3} \propto (1-\rho^2)^2\, d\alpha_1\, d\alpha_2\, d\sigma_1\, d\sigma_2\, d\rho. \tag{6-9}$$

Now $J = \Sigma\Sigma\, g_{ik}\, d\alpha_i\, d\alpha_k$,

where $g_{ik} = E\left[-\dfrac{\partial^2}{\partial\alpha_i\, \partial\alpha_k} \log f(x, y, \alpha)\right]$.

For the distribution under consideration the matrix $[g_{ik}]$ is

$$\begin{bmatrix} \dfrac{1}{\sigma_1^2(1-\rho^2)} & -\dfrac{\rho}{\sigma_1\sigma_2(1-\rho^2)} & 0 & 0 & 0 \\ -\dfrac{\rho}{\sigma_1\sigma_2(1-\rho^2)} & \dfrac{1}{\sigma_2^2(1-\rho^2)} & 0 & 0 & 0 \\ 0 & 0 & \dfrac{2-\rho^2}{\sigma_1^2(1-\rho^2)} & -\dfrac{\rho^2}{\sigma_1\sigma_2(1-\rho^2)} & -\dfrac{\rho}{\sigma_1(1-\rho^2)} \\ 0 & 0 & -\dfrac{\rho^2}{\sigma_1\sigma_2(1-\rho^2)} & \dfrac{2-\rho^2}{\sigma_2^2(1-\rho^2)} & -\dfrac{\rho}{\sigma_2(1-\rho^2)} \\ 0 & 0 & -\dfrac{\rho}{\sigma_1(1-\rho^2)} & -\dfrac{\rho}{\sigma_2(1-\rho^2)} & \dfrac{1+\rho^2}{(1-\rho^2)^2} \end{bmatrix}$$

and $|g_{ik}| = 4/\sigma_1^4\sigma_2^4(1-\rho^2)^4$. Hence the invariant given by J is

$$|g_{ik}|^{1/2}\, d\alpha_1\, d\alpha_2\, d\sigma_1\, d\sigma_2\, d\rho \propto \frac{d\alpha_1\, d\alpha_2\, d\sigma_1\, d\sigma_2\, d\rho}{\sigma_1^2\sigma_2^2(1-\rho^2)^2}. \tag{6-10}$$

By Eq. (4-13) J when expressed in terms of the u's is

$$J = \Sigma\Sigma\, G_{ik}\, du_i\, du_k,$$

where $G_{ik} = -\dfrac{\partial^2 B}{\partial u_i\, \partial u_k}$.

$$\text{Hence } |G_{ik}|^{1/2}\, du_1\, du_2\, du_3\, du_4\, du_5 = |g_{ik}|^{1/2}\, d\alpha_1\, d\alpha_2\, d\sigma_1\, d\sigma_2\, d\rho$$

$$\propto \frac{d\alpha_1\, d\alpha_2\, d\sigma_1\, d\sigma_2\, d\rho}{\sigma_1^2\sigma_2^2(1-\rho^2)^2}. \tag{6-11}$$

Multiplying the invariants (6-9) and (6-11) we shall have the new invariant

$$\frac{|G_{ik}|^{1/2}\, du_1^2\, du_2^2\, du_3^2\, du_4^2\, du_5^2}{u_1^3 u_2^3} \propto \frac{d\alpha_1^2\, d\alpha_2^2\, d\sigma_1^2\, d\sigma_2^2\, d\rho^2}{\sigma_1^2\sigma_2^2},$$

or

$$\frac{|G_{ik}|^{1/4}\, du_1\, du_2\, du_3\, du_4\, du_5}{u_1^{3/2}u_2^{3/2}} \propto \frac{d\alpha_1\, d\alpha_2\, d\sigma_1\, d\sigma_2\, d\rho}{\sigma_1\sigma_2}. \tag{6-12}$$

Hence the invariant $|G_{ik}|^{1/4}\, du_1\, du_2\, du_3\, du_4\, du_5/u_1^{3/2}u_2^{3/2}$ leads to the desired prior-probability form.

It appears that it may be possible to express in like manner the prior-probability form for the most general multivariate normal distribution as an invariant with the combination of the u's and J, but then the problem will be considerably cumbrous.

6.6.6 The multinomial distribution

Suppose that the population consists of p different types and the chances of the respective types occurring at any trial are $\alpha_1, \alpha_2, \ldots, \alpha_p$ where $\sum_{j=1}^{p} \alpha_j = 1$. If $x_1, x_2, \ldots, x_p$ are the numbers of the various types in a sample of n observations (where $\sum_{q=1}^{p} x_q = n$), then the probability of the sample is

$$f(x_q, \alpha_j) = \frac{n!}{x_1!x_2! \ldots x_p!} \alpha_1^{x_1} \alpha_2^{x_2} \ldots \alpha_p^{x_p}$$

$$= \exp\left\{\sum_{k=1}^{p} x_k \log \alpha_k - \log \prod_{q=1}^{p} x_q! + \log n!\right\}.$$

Only p-1 of the parameters α_j are independent, and we can take them to be $\alpha_1, \alpha_2, \ldots, \alpha_{p-1}$. Writing

$$\alpha_p = 1 - \sum_{k=1}^{p-1} \alpha_k,$$

$$f(x, \alpha_j) = \exp\left[\sum_{k=1}^{p-1} x_k \log \alpha_k + \left(n - \sum_{k=1}^{p-1} x_k\right) \log\left(1 - \sum_{k=1}^{p-1} \alpha_k\right)\right.$$

$$\left. - \log \prod_{q=1}^{n} x_q! + \log n!\right]$$

$$= \exp\left\{\sum_{k=1}^{p-1} x_k \log \frac{\alpha_k}{1 - \sum_{k=1}^{p-1} \alpha_k} + \ldots\right\}.$$

Hence $u_k = \log \dfrac{\alpha_k}{1 - \sum_{k=1}^{p-1} \alpha_k}$.

We note that $\dfrac{\partial u_k}{\partial \alpha_i} = \dfrac{1}{1 - \sum_{k=1}^{p-1} \alpha_k} = \dfrac{1}{\alpha_p}$ $(i \neq k)$,

and $$\frac{\partial u_k}{\partial \alpha_k} = \frac{1}{\alpha_k} + \frac{1}{\alpha_p}.$$

$$\frac{\partial(u_1, u_2, \ldots, u_{p-1})}{\partial(\alpha_1, \alpha_2, \ldots, \alpha_{p-1})} = \begin{vmatrix} \frac{1}{\alpha_1} + \frac{1}{\alpha_p} & \frac{1}{\alpha_p} & \cdots & \frac{1}{\alpha_p} \\ \frac{1}{\alpha_p} & \frac{1}{\alpha_2} + \frac{1}{\alpha_p} & \cdots & \frac{1}{\alpha_p} \\ \cdots & \cdots & \cdots & \cdots \\ \frac{1}{\alpha_p} & \frac{1}{\alpha_p} & \cdots & \frac{1}{\alpha_{p-1}} + \frac{1}{\alpha_p} \end{vmatrix}$$

$$= \frac{1}{\alpha_1 \alpha_2 \ldots \alpha_p}.$$

Thus $du_1\ du_2 \ldots du_{p-1} = \dfrac{d\alpha_1\ d\alpha_2 \ldots d\alpha_{p-1}}{\alpha_1\ \alpha_2 \ldots \alpha_p}$. (6-13)

Since each u_k has the range $(-\infty, \infty)$, $du_1\ du_2 \ldots du_{p-1}$ is a differential invariant of Type II. The appropriate prior-probability form is $d\alpha_1\ d\alpha_2 \ldots d\alpha_{p-1}$, but we cannot express it as an invariant in terms of the u's. We now take the help of J. The invariant given by J is (cf. Jeffreys, 1948, p. 162)

$$|G_{ik}|^{1/2}\ du_1\ du_2 \ldots du_{p-1} = |g_{ik}|^{1/2}\ d\alpha_1\ d\alpha_2 \ldots d\alpha_{p-1}$$

$$= \frac{d\alpha_1\ d\alpha_2 \ldots d\alpha_{p-1}}{(\alpha_1\ \alpha_2 \ldots \alpha_p)^{1/2}},$$

or

$$|G_{ik}|\ du_1^2\ du_2^2 \ldots du_{p-1}^2 = \frac{d\alpha_1^2\ d\alpha_2^2 \ldots d\alpha_{p-1}^2}{\alpha_1 \alpha_2 \ldots \alpha_p}. \qquad (6\text{-}14)$$

Dividing the invariant (6-14) by the invariant (6-13) we shall have the new invariant

$$|G_{ik}|\ du_1\ du_2 \ldots du_{p-1} = d\alpha_1\ d\alpha_2 \ldots d\alpha_{p-1} \qquad (6\text{-}15)$$

which leads to the desired prior-probability form. The case when the parameters are a set of chances can be treated on similar lines.

6.7 Distributions Admitting an Augmented Sufficient Set of Statistics

For distributions admitting an augmented sufficient set of statistics, we consider two examples.

6.7.1 The correlation coefficient

The one-parameter bivariate normal distribution, referred to in Sec. 2.4, is

$$f(x, y, \rho) = \frac{1}{2\pi\sigma_1\sigma_2(1-\rho^2)^{1/2}} \exp\left[- \frac{1}{2(1-\rho^2)} \left(\frac{x^2}{\sigma_1^2} - \frac{2\rho xy}{\sigma_1\sigma_2} + \frac{y^2}{\sigma_2^2} \right) \right], \tag{6-16}$$

where σ_1 and σ_2 are known. This case is interesting and is an instance where some caution must be exercised before writing the distribution in the standard form with standard ranges. At first we might think that the standard form is

$$f(x, y, \rho) = \exp\left[- \frac{x^2}{2\sigma_1^2(1-\rho^2)} - \frac{y^2}{2\sigma_2^2(1-\rho^2)} + \frac{\rho xy}{\sigma_1\sigma_2(1-\rho^2)} - \log 2\pi\sigma_1\sigma_2(1-\rho^2)^{1/2} \right], \tag{6-17}$$

where the number of u's is three. But since σ_1 and σ_2 are known, the two terms containing x^2 and y^2 can be now combined into a single term, thus reducing the number of u's to two. It is an essential condition of the standard form that the number of u's must be minimum. The standard form is therefore

$$f(x, y, \rho) = \exp\left[- \frac{1}{2(1-\rho^2)} \left(\frac{x^2}{\sigma_1^2} + \frac{y^2}{\sigma_2^2} \right) + \frac{\rho xy}{\sigma_1\sigma_2(1-\rho^2)} - \log 2\pi\sigma_1\sigma_2(1-\rho^2)^{1/2} \right]. \tag{6-18}$$

But in this form $u_1 = 1/(1-\rho^2)$ has the range $(1, \infty)$, which is not a standard range. To write $f(x, y, \rho)$ in the standard form with standard ranges,

$$f(x, y, \rho) = \exp\left[-\frac{1}{2}\left(\frac{1}{1-\rho^2} - 1\right)\left(\frac{x^2}{\sigma_1^2} + \frac{y^2}{\sigma_2^2}\right) - \frac{1}{2}\left(\frac{x^2}{\sigma_1^2} + \frac{y^2}{\sigma_2^2}\right) + \frac{\rho xy}{\sigma_1\sigma_2(1-\rho^2)} - \log 2\pi\sigma_1\sigma_2(1-\rho^2)^{1/2}\right]$$

$$= \exp\left[-\frac{\rho^2}{2(1-\rho^2)}\left(\frac{x^2}{\sigma_1^2} + \frac{y^2}{\sigma_2^2}\right) - \frac{1}{2}\left(\frac{x^2}{\sigma_1^2} + \frac{y^2}{\sigma_2^2}\right) + \frac{\rho xy}{\sigma_1\sigma_2(1-\rho^2)} - \log 2\pi\sigma_1\sigma_2(1-\rho^2)^{1/2}\right].$$

Now $u_1 = \rho^2/(1-\rho^2)$ has the standard range $(0, \infty)$, and $u_2 = \rho/(1-\rho^2)$ has the standard range $(-\infty, \infty)$. The appropriate prior-probability form for the correlation coefficient is $d\rho$. Now

$$du_1 \propto \frac{\rho}{(1-\rho^2)^2}\, d\rho, \qquad du_2 \propto \frac{1+\rho^2}{(1-\rho^2)^2}\, d\rho,$$

$$\frac{du_2}{du_1} \propto \frac{1+\rho^2}{\rho}, \qquad d\left(\frac{du_2}{du_1}\right) \propto \frac{1-\rho^2}{\rho^2}\, d\rho.$$

Finally, $u_1\, d\left(\frac{du_2}{du_1}\right) \propto d\rho$.

It is easy to show that $u_1\, d(du_2/du_1)$ is an invariant. du_1 and du_2 are differential invariants of Type II. du_2/du_1 is therefore an invariant of Type II. $d(du_2/du_1)$ is a differential invariant of the same type, and so is $u_1\, d(du_2/du_1)$.

It may be remarked also that u_1 and u_2 can be distinguished from each other by their different ranges. Hence the invariant $u_1\, d(du_2/du_1)$ leads to the uniform assessment of the prior-probability of the correlation coefficient.

6.7.2 Weibull Distribution

Consider the distribution referred to by Edgeworth (1908),

$$f(x, \alpha) = \frac{2}{\Gamma(1/4)} \exp[-(x-\alpha)^4] \quad (-\infty < x < \infty).$$

When written in the standard form with standard ranges,

$$f(x, \alpha) = \exp[4\alpha x^3 - 6\alpha^2 x^2 + 4\alpha^3 x - x^4 - \alpha^4 + \log 2 - \log \Gamma(1/4)].$$

$u_1 = \alpha$ and $u_3 = \alpha^3$ have each the standard range $(-\infty, \infty)$, and $u_2 = \alpha^2$ has the standard range $(0, \infty)$. Since α is a location paramater, its prior-probability distribution should be uniform.

$$\frac{du_2}{\sqrt{u_2}} \propto d\alpha.$$

Further $du_2/\sqrt{u_2}$ is an invariant. u_2 is easily identified, being the only u having the range $(0, \infty)$.

7. Invariants of Distributions Admitting Sufficient Statistics: 3. The Range Depending on the Parameter

7.1 Introduction

In this section the invariance theory is extended to the case when the range of the distribution depends on the parameter. More generally, we shall take up distributions admitting an augmented sufficient pair of statistics which, as in Eq. (2-6), are given by

$$f(x, \alpha) = \frac{g(x)}{h(\alpha)} \quad [a(\alpha) \le x \le b(\alpha)]. \tag{7-1}$$

We shall divide the distributions given by Eq. (7-1) into two classes: Class I, for which $h(\alpha)$ is not a numerical constant (independent of the parameter α); and Class II, for which $h(\alpha)$ is a numerical constant. For a distribution of Class II we have

$$f(x, \alpha) = g(x) \quad [a(\alpha) \le x \le b(\alpha)]. \tag{7-2}$$

The rectangular distribution in which the unknown parameter is the center, is an instance of a Class II distribution in which $g(x)$ is also a constant.

For distributions of Class I two sets of invariants are available: invariants based on the p.d.f., i.e., on $h(\alpha)$; and invariants based on the range (a, b). For distributions of Class II, only the range (a, b) supplies us invariants.

7.2 Distributions of Class I [Invariants Based on $h(\alpha)$ only]

The form (7-1) will be said to be the standard form of a distribution of Class I. When a distribution is written in the standard form, $h(\alpha)$ is unique except for a multiplicative constant. For we can write

$$f(x, \alpha) = \frac{g(x)}{h(\alpha)} = \frac{\lambda_1 \, g(x)}{\lambda_1 \, h(\alpha)},$$

so that any multiplicative constant λ_1 can be associated with $g(x)$ and $h(\alpha)$. Further, since $g(x)/h(\alpha)$ is positive for all values of x and α, both $g(x)$ and $h(\alpha)$ are of the same sign for all values of x and α. The constant λ_1 at our disposal enables us to render both $g(x)$ and $h(\alpha)$ positive for all x and α. Hence it will be understood that $h(\alpha)$ is positive for all α. It follows that if $f(x, \alpha)$ is written in the standard form in any two ways such as $g(x)/h(\alpha)$ and $g_1(x)/h_1(\alpha)$, then $h_1(\alpha) = \lambda h(\alpha)$, where λ is a *positive* constant.

7.2.1 Standard ranges

Since $h(\alpha)$ is positive, we define the following three ranges as standard ranges for $h(\alpha)$: $(0, \infty)$, $(0, 1)$, and $(1, k)$, where k is any number greater than unity, which may be $+\infty$. If $f(x, \alpha)$ is written in the standard form such that $h(\alpha)$ has a standard range, it will be said to be written in the "standard form with standard range."

We can always write $f(x, \alpha)$ in the standard form with standard range. For in choosing $h(\alpha)$ we have a positive multiplicative constant at our choice. If $h(\alpha)$ has the range $(0, \infty)$, it has already a standard range. If it has the range (0, m), then $h_1(\alpha) = (1/m)\ h(\alpha)$ has the standard range (0, 1). If it has the range (ℓ, m), then $h_1(\alpha) = (1/\ell)\ h(\alpha)$ has the standard range (1, k), where $k = m/\ell > 1$.

LEMMA 1

If $f(x, \alpha)$ is written in the standard form with standard range in any two ways giving $h(\alpha)$ and $h_1(\alpha)$ respectively, then

$$h_1(\alpha) = \lambda\ h(\alpha) \quad \text{if the standard range is } (0, \infty),$$

$$h_1(\alpha) = h(\alpha) \quad \text{if the standard range is } (0, 1) \text{ or } (1, k).$$

The proof is very simple.

7.2.2 Transformations

Consider any transformation $x = \phi(x')$, $\alpha = \psi(\alpha')$. The p.d.f. of x' is

$$f'(x', \alpha') = \frac{g[\phi(x')]}{h[\psi(\alpha')]} \left|\frac{d\phi}{dx'}\right| .$$

Now $|d\phi/dx'|$ is a function of x' only. Hence, if $f'(x', \alpha')$ is written in the standard form in any way as $g'(x')/h'(\alpha')$, we shall have

$$h'(\alpha') = \lambda\ h[\psi(\alpha')] = \lambda\ h(\alpha)$$

for corresponding values of α and α'. If both $f(x, \alpha)$ and $f'(x', \alpha')$ are written in the standard form with standard range, we shall have, for corresponding values of α and α', $h'(\alpha') = \lambda h(\alpha)$ if the standard range is $(0, \infty)$, and $h'(\alpha') = h(\alpha)$ if the standard range is (0, 1) or (1, k).

7.2.3 Invariants

The results of the preceding section may be stated as follows: When a distribution of Class I is always written in the standard form with standard range, for all transformations of the parameter and the variate

1. $h(\alpha)$ is an invariant of Type II if its standard range is $(0, \infty)$. $h^{\ell}\ dh$ is a differential invariant of Type II, where ℓ is any constant.

2. $h(\alpha)$ is an invariant of Type I if its standard range is (0, 1) or (1, k). $\phi(h)\,dh$ is a differential invariant of Type I, where ϕ is any arbitrary function.

7.2.4 Applications to the prior probability

Invariance rule. It is possible to state an invariance rule for the distributions of Class I. In all cases dh/h is an invariant, so that we may take the prior probability of dh to be proportional to dh/h. This rule is found to be satisfactory for practical distributions. There is of course choice for its modification in cases where it is not satisfactory.

Example 1. For the rectangular distribution

$$f(x, \sigma) = \frac{1}{2\sigma} \quad (-\sigma \leq x \leq \sigma),$$

$$h(\sigma) = \frac{1}{2\sigma} \quad \text{and} \quad \frac{dh}{h} \propto \frac{d\sigma}{\sigma}.$$

Hence the rule leads to the prior-probability form $d\sigma/\sigma$, which is quite satisfactory, σ being a scale parameter.

Example 2. Consider the distribution

$$f(x, \alpha) = e^{-(x-\alpha)} \quad (\alpha \leq x < \infty).$$

Here $f(x, \alpha) = e^{-x}/e^{-\alpha}$, and $h(\alpha) = e^{-\alpha}$ $dh/h \propto d\alpha$, so that the prior probability of α is uniform by the rule. This is also satisfactory, α being a location parameter.

Example 3. Take Pearson's Type XI distribution,

$$f(x, \alpha) = (m-1)\alpha^{m-1}x^{-m} \quad (\alpha \leq x < \infty,\ m > 1),$$

where m is a known constant. $h(\alpha) = 1/\alpha^{m-1}$. $dh/h \propto d\alpha/\alpha$, which is again the usual form, α being now a scale parameter.

7.3 Distributions of Class I [Invariants Based on the Range (a, b)]

It will be shown that we can obtain a second set of invariants by reducing a distribution of Class I to a simpler form.

LEMMA 1

Every distribution of Class I can be reduced, by a transformation of x only, to the rectangular form

$$F(X, \alpha) = \frac{1}{H(\alpha)} \quad [A(\alpha) \le X \le B(\alpha)].$$

Proof. The only transformation of x which will transform Eq. (7-1) into the rectangular form is $dX = \lambda\ g(x)\ dx$, where λ is any positive constant, or

$$X = \lambda\ \phi(x) + \mu, \tag{7-3}$$

where $\phi(x) = \int g(x)\ dx$, and μ is another constant at our disposal. And the transformation (7-3) transforms the given distribution into

$$F(X, \alpha) = \frac{1}{\lambda\ h(\alpha)} \quad [\lambda\ \phi(a) + \mu \le X \le \lambda\ \phi(b) + \mu],$$

that is, into

$$F(X, \alpha) = \frac{1}{H(\alpha)} \quad [A(\alpha) \le X \le B(\alpha)], \tag{7-4}$$

where $H(\alpha) = \lambda\ h(\alpha)$, $A(\alpha) = \lambda\ \phi(a) + \mu$, $B(\alpha) = \lambda\ \phi(b) + \mu$. (7-5)

Since $\int_A^B F(X, \alpha)\ dX \equiv 1$,

we also have

$$B(\alpha) - A(\alpha) \equiv H(\alpha).$$

We note that the transformation (7-3) is unique except for a positive multiplicative constant λ and an additive constant μ, which are at our choice. The p.d.f. $H(\alpha)$ is unique except for the multiplicative constant λ, and the extremities $A(\alpha)$ and $B(\alpha)$ of the range of the reduced distribution are unique except for the multiplicative and additive constants λ and μ.

Definitions. The reduced rectangular form (7-4) for a Class I distribution will be called the "reduced standard form of a Class I distribution." The transformation (7-3) will be called a "reduction transformation" in order to distinguish it from other arbitrary transformations of the parameter and the variate.

It follows that if any two reduction transformations transform a Class I distribution into reduced standard forms

$$F(X, \alpha) = \frac{1}{H(\alpha)} \quad [A(\alpha) \le X \le B(\alpha)],$$

and $$F_1(X, \alpha) = \frac{1}{H_1(\alpha)} = [A_1(\alpha) \le X \le B_1(\alpha)],$$

respectively, then

$$H_1(\alpha) = \lambda_1 H(\alpha), \qquad A_1(\alpha) = \lambda_1 A(\alpha) + \mu_1,$$

$$B_1(\alpha) = \lambda_1 B(\alpha) + \mu_1,$$

where λ_1 and μ_1 are constants.

We shall now use the device of standard ranges to fix the choice constants λ and μ in the reduced standard form. Since $B(\alpha) - A(\alpha) \equiv H(\alpha)$, $B(\alpha)$ will not lead to any new invariants other than those given by $A(\alpha)$ and $H(\alpha)$. Accordingly our interest will center only on $A(\alpha)$ and $H(\alpha)$, for which we shall define standard ranges.

7.3.1 Standard ranges

The procedure of defining standard ranges for $A(\alpha)$ and $H(\alpha)$ is slightly complicated.

If the range of $A(\alpha)$ is limited in both directions, take $(-1, 1)$ as the standard range of $A(\alpha)$, and it is not necessary then to define a standard range for $H(\alpha)$.

If the range of $A(\alpha)$ is not limited in both directions, take the same standard ranges as in Sec. 7.2.1 for $H(\alpha)$, viz. $(0, \infty)$, $(0, 1)$, and $(1, k)$. Take the three ranges $(-\infty, \infty)$, $(-\infty, 0)$ and $(0, \infty)$ as the standard ranges for $A(\alpha)$.

If a distribution of Class I is reduced to the standard form (7-4) such that $A(\alpha)$ and $H(\alpha)$ have standard ranges, it will be said to be reduced to the "reduced standard form with standard ranges."

We can always find a transformation for a Class I distribution leading to the reduced standard form with standard ranges. For in choosing a reduction transformation (7-3) there are two constants λ and μ at our disposal. λ and μ are uniquely determined when the standard range of $A(\alpha)$ is $(-1, 1)$. For if two choices give $H(\alpha)$, $A(\alpha)$ and $H_1(\alpha)$, $A_1(\alpha)$ respectively, then we have

$$H_1(\alpha) = \lambda_1 H(\alpha) \quad \text{and} \quad A_1(\alpha) = \lambda_1 A(\alpha) + \mu_1.$$

As $A(\alpha)$ and $A_1(\alpha)$ both have the range $(-1, 1)$ we must have $\lambda_1 = \pm 1$ and $\mu_1 = 0$. But since λ_1 is necessarily positive, we must have $\lambda_1 = +1$. Hence $H_1(\alpha) = H(\alpha)$ and $A_1(\alpha) = A(\alpha)$. It is not necessary in this case to define a standard range for $H(\alpha)$, which is uniquely determined.

If the range of $A(\alpha)$ is not limited in both directions, choose the constant λ in the reduction transformation so that $H(\alpha)$ has one of the standard ranges $(0, \infty)$, $(0, 1)$, or $(1, k)$. λ remains undecided only when $H(\alpha)$ has the range $(0, \infty)$; we can then choose λ and μ so that $A(\alpha)$ has one of the standard ranges $(-\infty, \infty)$, $(-\infty, 0)$, or $(0, \infty)$. μ is determined only if the standard range of $A(\alpha)$ is $(-\infty, \infty)$ or $(0, \infty)$; otherwise both λ and μ still remain undecided.

When $H(\alpha)$ has the standard range $(0, 1)$ or $(1, k)$, λ is uniquely determined and we still have the choice to determine μ. In this case $A_1(\alpha) = A(\alpha) + \mu_1$, where A_1 and A correspond to two different choices of μ. If the standard range is $(-\infty, 0)$ or $(0, \infty)$, $\mu_1 = 0$; otherwise μ_1 remains undecided. Consequently μ_1 is uniquely determined only if the standard range of $A(\alpha)$ is $(-\infty, 0)$ or $(0, \infty)$. Hence we have the following Lemma.

LEMMA 2

If any two transformations give the reduced standard forms with standard ranges

$$F(X, \alpha) = \frac{1}{H(\alpha)} \quad [A(\alpha) \le X \le B(\alpha)],$$

and $$F_1(X, \alpha) = \frac{1}{H_1(\alpha)} \quad [A_1(\alpha) \le X \le B_1(\alpha)],$$

then

1. $H_1(\alpha) = H(\alpha)$ and $A_1(\alpha) = A(\alpha)$ if the standard range of $A(\alpha)$ is $(-1, 1)$.

2. $H_1(\alpha) = \lambda_1 H(\alpha)$ and $A_1(\alpha) = \lambda_1 A(\alpha) + \mu$ if the standard range of $H(\alpha)$ is $(0, \infty)$ and that of $A(\alpha)$ is $(-\infty, \infty)$.

3. $H_1(\alpha) = \lambda_1 H(\alpha)$ and $A_1(\alpha) = \lambda_1 A(\alpha)$ if the standard range of $H(\alpha)$ is $(0, \infty)$ and that of $A(\alpha)$ is $(0, \infty)$ or $(-\infty, 0)$.

4. $H_1(\alpha) = H(\alpha)$ and $A_1(\alpha) = A(\alpha) + \mu_1$ if the standard range of $H(\alpha)$ is $(0, 1)$ or $(1, k)$ and that of $A(\alpha)$ is $(-\infty, \infty)$.

5. $H_1(\alpha) = H(\alpha)$ and $A_1(\alpha) = A(\alpha)$ if the standard range of $H(\alpha)$ is $(0, 1)$ or $(1, k)$ and that of $A(\alpha)$ is $(0, \infty)$ or $(-\infty, 0)$.

7.3.2 Transformations

Now consider any general transformation $x = \xi(x')$, $\alpha = \eta(\alpha')$. This transforms a Class I distribution into

$$f'(x', \alpha') = \frac{g[\xi(x')]}{h[\eta(\alpha')]} \left|\frac{d\xi}{dx'}\right| \quad \{a[\eta(\alpha')] \leq \xi(x') \leq b[\eta(\alpha')]\}. \tag{7-6}$$

Let us obtain the reduced standard form of Eq. (7-6). The reduction transformation is

$$X' = \lambda' \int g[\xi(x')] \left|\frac{d\xi}{dx'}\right| dx' + \mu'$$

$$= \lambda' \int g(\xi)\, d\xi + \mu',$$

or $$X' = \lambda' \phi(\xi) + \mu', \tag{7-7}$$

where $\phi(x) = \int g(x)\, dx$ as in Sec. 7.3. And from Eqs. (7-6) and (7-7) the limits of X' are

$$\lambda' \phi\{a[\eta(\alpha')]\} + \mu' \leq X' \leq \lambda' \phi\{b[\eta(\alpha')]\} + \mu'.$$

Hence the reduced standard form of Eq. (7-6) is

$$F'(X', \alpha') = \frac{1}{H'(\alpha')} \quad [A'(\alpha') \leq X' \leq B'(\alpha')], \tag{7-8}$$

where $H'(\alpha') = \lambda' h[\eta(\alpha')]$, $A'(\alpha') = \lambda' \phi\{a[\eta(\alpha')]\} + \mu'$,

$$B'(\alpha') = \lambda' \phi\{b[\eta(\alpha')]\} + \mu'. \tag{7-9}$$

Now $h[\eta(\alpha')] = h(\alpha)$, $\phi\{a[\eta(\alpha')] = \phi(a)$, $\phi\{b[\eta(\alpha')]\} = \phi(b)$.

Hence, comparing Eq. (7-9) with Eq. (7-5), we have for corresponding values of α and α'

$$H'(\alpha') = \lambda_1 H(\alpha), \quad A'(\alpha') = \lambda_1 A(\alpha) + \mu_1, \quad B'(\alpha') = \lambda_1 B(\alpha) + \mu_1. \tag{7-10}$$

The constants λ_1 and μ_1 in Eq. (7-10) may be determined in some cases by the device of standard ranges as in Sec. 7.3. This leads us to the following lemma.

LEMMA 3

When a distribution of Class I is written in the reduced standard form with standard ranges, *both before and after* any arbitrary transformation as

$$F(X, \alpha) = \frac{1}{H(\alpha)} \quad [A(\alpha) \leq X \leq B(\alpha)],$$

and

$$F'(X', \alpha') = \frac{1}{H'(\alpha')} \quad [A'(\alpha') \leq X' \leq B'(\alpha')],$$

respectively, then for corresponding values of α and α' we have:

1. $H'(\alpha') = H(\alpha)$ and $A'(\alpha') = A(\alpha)$ if the standard range of $A(\alpha)$ is $(-1, 1)$.
2. $H'(\alpha') = \lambda_1 H(\alpha)$ and $A'(\alpha') = \lambda_1 A(\alpha) + \mu_1$ if the standard range of $H(\alpha)$ is $(0, \infty)$ and that of $A(\alpha)$ is $(-\infty, \infty)$.
3. $H'(\alpha') = \lambda_1 H(\alpha)$ and $A'(\alpha') = \lambda_1 A(\alpha)$ if the standard range of $H(\alpha)$ is $(0, \infty)$ and that of $A(\alpha)$ is $(0, \infty)$ or $(-\infty, 0)$.
4. $H'(\alpha') = H(\alpha)$ and $A'(\alpha') = A(\alpha) + \mu_1$ if the standard range of $H(\alpha)$ is $(0, 1)$ or $(1, k)$ and that of $A(\alpha)$ is $(-\infty, \infty)$.
5. $H'(\alpha') = H(\alpha)$ and $A'(\alpha') = A(\alpha)$ if the standard range of $H(\alpha)$ is $(0, 1)$ or $(1, k)$ and that of $A(\alpha)$ is $(0, \infty)$ or $(-\infty, 0)$.

7.2.3 Invariants

We now restate the results of the preceding lemma. When a distribution of Class I is always written in the reduced standard form with standard ranges, for all transformations of the parameter and the variate we have:

1. $H(\alpha)$ and $A(\alpha)$ are invariants of Type I if the standard range of $A(\alpha)$ is $(-1, 1)$. $d\,\phi(H)$ and $d\,\psi(A)$ are differential invariants of Type I, ϕ and ψ being any arbitrary functions.

2. $H(\alpha)$ is an invariant of Type II and $A(\alpha)$ of Type III if the standard range of $H(\alpha)$ is $(0, \infty)$ and that of $A(\alpha)$ is $(-\infty, \infty)$. $H^{\ell}\,dH$ and dA are differential invariants of Type II, ℓ being any constant.

3. Both $H(\alpha)$ and $A(\alpha)$ are invariants of Type II if the standard range of $H(\alpha)$ is $(0, \infty)$ and that of $A(\alpha)$ is $(0, \infty)$ or $(-\infty, \infty)$. $H^{\ell}\,dH$ and $A^m\,dA$ are differential invariants of Type II.

4. $H(\alpha)$ and $A(\alpha)$ are invariants of Type I and Type II[1] respectively if the standard range of $H(\alpha)$ is $(0, 1)$ or $(1, k)$ and that of $A(\alpha)$ is $(-\infty, \infty)$. $d\,\phi(H)$ and dA are differential invariants of Type I.

5. Both $H(\alpha)$ and $A(\alpha)$ are invariants of Type I if the standard range of $H(\alpha)$ is $(0, 1)$ or $(1, k)$ and that of $A(\alpha)$ is $(0, \infty)$ or $(-\infty, 0)$. $d\,\phi(H)$ and $d\,\psi(A)$ are differential invariants of Type I.

7.3.4 Combination of invariants

We can obtain more invariants by combining the invariants given by $H(\alpha)$ with those given by $A(\alpha)$. Thus in all cases dH and dA are differential invariants of Type II and the constant of proportionality λ is the same for both. Hence dH/dA is an invariant of Type I. Consequently $\phi(dH/dA)$ and $d\,\phi(dH/dA)$ are invariants of Type I, ϕ being any arbitrary function.

7.3.5 Applications to the prior probability

It is not necessary to consider applications of this second set of invariants, since the first set and the invariance rule given in Sec. 7.2.4 are quite sufficient for the purposes of the prior probability. We shall take only one example to illustrate the calculation of the invariants of the second set.

For the distribution

$$f(x, \alpha) = e^{-(x-\alpha)} \qquad (\alpha \leq x < \infty),$$

the reduction transformation is

$$X = \lambda \int e^{-x}\, dx + \mu = -\lambda\, e^{-x} + \mu.$$

The reduced standard form is

$$F(X, \alpha) = \frac{1}{\lambda\, e^{-\alpha}}; \quad (-\lambda\, e^{-\alpha} + \mu \leq X \leq \mu).$$

Here $H(\alpha) = \lambda\, e^{-\alpha}$, $A(\alpha) = -\lambda\, e^{-\alpha} + \mu$.

Since the range of α is $(-\infty, \infty)$, the range of $H(\alpha)$ is $(0, \infty)$ and that of $A(\alpha)$ is $(-\infty, \mu)$. If we choose $\mu = 0$, $A(\alpha)$ will have the standard range $(-\infty, 0)$. Hence

$$H(\alpha) = \lambda\, e^{-\alpha}, \quad A(\alpha) = -\lambda\, e^{-\alpha}.$$

dH/H and dA/A are both invariants and lead to the same prior-probability form $d\alpha$. The invariant dH/dA of Sec. 7.3.4 reduces in this case to the numerical constant -1.

7.4 Distributions of Class II

Let $f(x, \alpha) = g(x) \quad [a(\alpha) \leq x \leq b(\alpha)]$. (7-11)

As remarked in Sec. 7.1, only the range (a, b) can now give us invariants. As in Sec. 7.3, we shall reduce Eq. (7-11) to a simpler form.

7.4.1 Transformations

Consider the transformation

$$X = \int g(x)\, dx + \mu,$$

where μ is any constant, or

$$X = \phi(x) + \mu, \tag{7-12}$$

where $\phi(x) = \int g(x)\, dx$. The transformation (7-12) reduces Eq. (7-11) to the form

$$F(X, \alpha) = 1 \quad [A(\alpha) \leq X \leq B(\alpha)], \tag{7-13}$$

where $A(\alpha) = \phi(a) + \mu$, $B(\alpha) = \phi(b) + \mu$, $B(\alpha) - A(\alpha) \equiv 1$. (7-14)

The form (7-13) will be called the "reduced standard form of a Class

II distribution." The reduced standard form is unique except for an additive constant μ in the extremities $A(\alpha)$ and $B(\alpha)$. The reduction transformation (7-12) is unique except for the additive constant μ. In virtue of Eq. (7-14), $B(\alpha)$ will not give any new invariants other than those given by $A(\alpha)$. Hence our interest will lie only in $A(\alpha)$.

7.4.2 Standard ranges

It will be seen that the invariance theory of Class II distributions will be on similar lines to that of Class I distributions in Sec. 7.3. Whereas in the present case there is only one constant μ, there are two constants λ and μ in the previous case. The three ranges $(-\infty, \infty)$, $(-k, 0)$, and $(0, k)$, where k is any positive number which may be $+\infty$, will be taken as standard ranges for $A(\alpha)$. The form (7-13) will be called the "reduced standard form with standard range" if $A(\alpha)$ has a standard range.

7.4.3 Invariants

We shall now state (without proof) the invariants of Class II distributions, the proof being similar to that in the case of Class I distributions.

When a distribution of Class II is always written in the reduced standard form with standard range, for all transformations of the parameter and the variate,

1. $A(\alpha)$ is an invariant of Type II[1] and dA is a differential invariant of Type I if the standard range of $A(\alpha)$ is $(-\infty, \infty)$.

2. $A(\alpha)$ is an invariant of Type I if its standard range is $(-k, 0)$ or $(0, k)$. $d\,\phi(A)$ is a differential invariant of Type I, ϕ being an arbitrary function.

7.4.4 Applications to the prior probability

Invariance rule. Write a given distribution of Class II in the reduced standard form with standard range. Take the prior probability of dA to be proportional to dA or dA/A according as the standard range of A is or is not $(-\infty, \infty)$.

Example 1. For the rectangular distribution

$$f(x, \alpha) = \frac{1}{2} \quad (\alpha-1 \leq x \leq \alpha+1),$$

the reduction transformation is

$$X = \int \frac{1}{2}\,dx + \mu = \frac{1}{2}x + \mu,$$

and the reduced standard form is

$$F(X, \alpha) = 1 \qquad [\tfrac{1}{2}(\alpha-1) + \mu \leq X \leq \tfrac{1}{2}(\alpha+1) + \mu].$$

Here $A(\alpha) = \frac{1}{2}(\alpha-1) + \mu$ has the standard range $(-\infty, \infty)$, since the range of α is $(-\infty, \infty)$. $dA \propto d\alpha$, so that the rule gives uniform prior probability for α. This is satisfactory, α being a location parameter.

Example 2. Consider the distribution

$$f(x, \alpha) = 2x \qquad (\alpha \leq x \leq (\alpha^2 + 1)^{1/2},\ \alpha \text{ positive}).'$$

The reduction transformation is

$$X = \int 2x\,dx + \mu = x^2 + \mu,$$

and the reduced standard form is

$$F(X, \alpha) = 1 \qquad (\alpha^2 + \mu \leq X \leq \alpha^2 + \mu + 1).$$

Here $A(\alpha) = \alpha^2 + \mu$ has the range (μ, ∞). Choosing $\mu = 0$, $A(\alpha)$ will have the standard range $(0, \infty)$. Then $A(\alpha) = \alpha^2$ and $dA/A \propto d\alpha/\alpha$. The rule therefore gives the usual form $d\alpha/\alpha$, having the range $(0, \infty)$.

7.5 Distributions Depending on Two Parameters

The invariance theory for a single parameter developed in the previous sections can be extended to the case of two parameters. Consider the distributions of the type

$$f(x, \alpha, \beta) = \frac{g(x)}{h(\alpha, \beta)} \qquad [a(\alpha, \beta) \leq x \leq b(\alpha, \beta)]. \tag{7-15}$$

Now we do not get two classes of distributions as we did in the case of a single parameter, according as $h(\alpha, \beta)$ is or is not a numerical constant (independent of the parameters). For it is easily seen that $h(\alpha, \beta)$ cannot be a numerical constant. If $h(\alpha, \beta)$ were a numerical constant k, we would have

$$\int_{a(\alpha,\ \beta)}^{b(\alpha,\ \beta)} g(x)\ dx \equiv k,$$

which means that the parameters α and β are not independent.

The theory of invariants of distributions (7-15) is exactly the same as that in Sec. 7.3. The reduced standard form for distributions (7-15) is

$$f(X,\ \alpha,\ \beta) = \frac{1}{H(\alpha,\ \beta)} \qquad [A(\alpha,\ \beta) \leq X \leq B(\alpha,\ \beta)]. \tag{7-16}$$

The invariance states in Secs. 7.3.2 and 7.3.4 hold good in the present case also.

7.5.1 Applications to the prior probability

Since we are now concerned with two parameters, we have to use invariants based on both $H(\alpha,\ \beta)$ and $A(\alpha,\ \beta)$.

Example 1. For the general rectangular distribution

$$f(x,\ \alpha,\ \sigma) = \frac{1}{2\sigma} \qquad (\alpha-\sigma \leq x \leq \alpha+\sigma),$$

the distribution is already in the reduced standard form. $H(\alpha,\ \sigma) = 1/2\sigma$ has the standard range $(0,\ \infty)$, and $A(\alpha,\ \sigma) = \alpha-\sigma$ has the standard range $(-\infty,\ \infty)$. dH/H and dA are differential invariants and so is $dH\ dA/H$. But $dH\ dA/H \propto d\alpha\ d\sigma/\sigma$, which is the appropriate prior-probability form for the rectangular distribution.

Example 2. Consider the distribution

$$f(x,\ \alpha,\ \beta) = \frac{1}{2(\beta-\alpha)\sqrt{x}} \qquad (\alpha \leq \sqrt{x} \leq \beta).$$

The reduction transformation is

$$X = \lambda x^{1/2} + \mu,$$

and the reduced standard form is

$$F(X,\ \alpha,\ \beta) = \frac{1}{\lambda(\beta-\alpha)} \qquad (\lambda\alpha + \mu \leq X \leq \lambda\beta + \mu).$$

Here $H(\alpha,\ \beta) = \lambda(\beta-\alpha)$ has the standard range $(0,\ \infty)$, and $A(\alpha,\ \beta) = \lambda\alpha + \mu$ has the standard range $(-\infty,\ \infty)$. $dH\ dA/H$ is a differential invariant giving the prior probability for $d\alpha\ d\beta/(\beta - \alpha)$.

8. Invariants of Distributions Admitting Sufficient Statistics: 4. An Alternative System of Invariants

8.1 Introduction

The basis of the invariance theory of the previous sections was the parametric functions selected from the p.d.f. of a distribution. Thus in Secs. 5 and 6 our system of invariants was constructed from the functions $u_k(\alpha_j)$. We now propose an alternative system of invariants based on the $v_k(x)$, that is, on the functions of the variate x. These new invariants enable us to arrive at an interesting invariance rule for the prior probability of a single parameter, which, for the first time, leads straight to the standard prior-probability forms for some well-known distributions.

The theory of this alternative system is similar to that of the previous one. Consequently we shall omit the proofs of the results.

8.2 Standard Form

We have

$$f(x, \alpha_j) = \exp\left[\sum_{k=1}^{\nu} u_k(\alpha_j)\ v_k(x) + A(x) + B(\alpha_j)\right]. \tag{8-1}$$

As before we shall call the form (8-1) the standard form of distributions under consideration. When a distribution is written in the standard form (8-1), the functions $v_k(x)$ are unique except for multiplicative and additive constants. In general we can replace v_k by $\lambda v_k + \mu$, where λ and μ are arbitrary constants.

Consider any transformation of x into x' and of the α_j into the α'_j. If a distribution (8-1) is written in the standard form both before and after any transformation, and if $v_k(x)$ and $v'_k(x')$ are corresponding functions before and after the transformation respectively, then for corresponding values of x and x'

$$v'_k(x') = \lambda\ v_k(x) + \mu, \tag{8-2}$$

where λ and μ are constants. Each $v_k(x)$ is therefore an invariant of Type III. If we define standard ranges for the v_k, as we did previously for the u_k, one or both of the constants λ and μ in Eq. (8-2) can be fixed up in some cases, thus reducing v_k to an invariant of Type II or I. However, we are not interested in the v_k's themselves. Being functions of x, they are not of use in stating the prior probability of the parameters. It will be useful if we can obtain from them some functions of the parameters that will be invariant. Such parametric functions can be obtained by taking the expectations of the $v_k(x)$'s. Taking expectations of both sides of Eq. (8-2), we have

$$E[v_k'(x')] = \lambda\, E[v_k(x)] + \mu,$$

or $$E_k' = \lambda\, E_k + \mu, \qquad (8\text{-}3)$$

where $E_k = E[v_k(x)]$ and $E_k' = E[v_k'(x')]$. Hence for all transformations of the variate and the parameters each E_k is an invariant of Type III. We now define standard ranges for the E_k, which may enable us to determine one or both of the constants in Eq. (8-3), thus reducing E_k to an invariant of Type II or I. As before we shall take the three ranges $(-\infty, \infty)$, $(0, \infty)$, and $(-1, 1)$ as the standard ranges. Since there is a choice of two constants in the selection of each $v_k(x)$, we can always choose the constants in such a way that $E_k = E[v_k(x)]$ has one of the three standard ranges. When a distribution is written in the standard form such that the E_k have standard ranges, it will be said to be written in the standard form with standard ranges.

8.3 Invariants

When a distribution is always written in the standard form with standard ranges, we have the following for all transformations of the parameters and the variate:

1. E_k is an invariant of Type III if its standard range is $(-\infty, \infty)$; dE_k is a differential invariant of Type II.

2. E_k is an invariant of Type II if its standard range is $(0, \infty)$; $E_k^{\ell}\, dE_k$ is a differential invariant of Type II, where ℓ is any constant.

3. $|E_k|$ is an invariant of Type I if the standard range of E_k is (-1, 1); $\phi(|E_k|)$ and d $\phi(|E_k|)$ are also invariants of Type I, where ϕ is any arbitrary function.

8.4 Applications to the Prior Probability

8.4.1 Single parameter

Let $f(x, \alpha) = \exp[u(\alpha)\ v(x) + A(x) + B(\alpha)]$, and let $E = E[v(x)]$. We now state a remarkable invariance rule as follows.

Invariance rule. Write the given distribution in the standard form with standard range. Take $P(dE|H) \propto dE$ if the standard range of E is $(-\infty, \infty)$ or (-1, 1), and $P(dE\ H) \propto dE/E$ if the standard range of E is $(0, \infty)$.

Let us see how this rule works for some well-known distributions.

8.4.2 The normal distribution

$$y = \frac{1}{\sigma\sqrt{2\pi}} \exp\left[-\frac{(x-\alpha)^2}{2\sigma^2}\right].$$

a. If σ is known, but α unknown,

$$f(x, \alpha) = \exp\left(\frac{\alpha x}{\sigma^2} - \frac{x^2}{2\sigma^2} - \frac{\alpha^2}{2\sigma^2} - \log \sigma\sqrt{2\pi}\right).$$

Here $v(x) = x$, and $E = E(x) = \alpha$ has the standard range $(-\infty, \infty)$. $dE = d\alpha$, so that the rule gives uniform prior probability for α.

b. If α is known but σ unknown,

$$f(x, \sigma) = \exp\left[-\frac{(x-\alpha)^2}{2\sigma^2} - \log \sigma\sqrt{2\pi}\right].$$

Now $v(x) = (x-\alpha)^2$, and $E = E(x-\alpha)^2 = \sigma^2$ has the standard range $(0, \infty)$. $dE/E \propto d\sigma/\sigma$, so that the rule gives the form $d\sigma/\sigma$.

8.4.3 The Poisson distribution

$$f(x, \alpha) = \frac{e^{-\alpha} x}{x!} = \exp(x \log \alpha - \log x! - \alpha).$$

$v(x) = x$, and $E = E(x) = \alpha$ has the standard range $(0, \infty)$. $dE/E = d\alpha/\alpha$, so that we have the prior-probability form $d\alpha/\alpha$.

8.4.4 The binomial distribution

$$f(x, \alpha) = \binom{n}{x} \alpha^x (1-\alpha)^{n-x}$$

$$= \exp\left[x \log\left(\frac{\alpha}{1-\alpha}\right) + \log\binom{n}{x} + n \log(1-\alpha)\right].$$

Here $v(x) = x$, and $E(x) = n\alpha$ has the range $(0, n)$, which is not a standard range. We rewrite $f(x, \alpha)$ as

$$f(x, \alpha) = \exp\left[\left(\frac{2}{n} x-1\right) \frac{n}{2} \log \frac{\alpha}{1-\alpha} + \log\binom{n}{x} + \frac{n}{2} \log\left(\frac{\alpha}{1-\alpha}\right)\right.$$
$$\left. + n \log(1-\alpha)\right].$$

Now $v(x) = (2/n)x-1$, and $E = E[(2/n)x-1] = 2\alpha-1$ has the standard range $(-1, 1)$. Then $dE \propto d\alpha$, and the rule gives uniform prior probability for α.

8.4.5 The case when the parameter α is a chance

Where x takes only the values 0 and 1,

$$f(x, \alpha) = \alpha^{1-x}(1-\alpha)^x$$

$$= \exp\left[(2x-1) \frac{1}{2} \log\left(\frac{1-\alpha}{\alpha}\right) + \frac{1}{2} \log\left(\frac{1-\alpha}{\alpha}\right) + \log \alpha\right].$$

$v(x) = 2x-1$, and $E = -2\alpha + 1$ has the standard range $(-1, 1)$. $dE \propto d\alpha$, and again we have uniform prior probability for α.

8.4.6 The correlation coefficient

$$f(x, y, \rho) = \frac{1}{2\pi(1-\rho^2)^{1/2}} \exp\left[-\frac{1}{2(1-\rho^2)} (x^2 - 2\rho xy + y^2)\right]$$

$$= \exp\left[-\frac{x^2 + y^2}{2(1-\rho^2)} + \frac{\rho xy}{1-\rho^2} - \log 2\pi(1-\rho^2)^{1/2}\right].$$

Here

$$v_1(x, y) = x^2 + y^2, \qquad v_2(x, y) = xy.$$

$$E_1 = E(x^2 + y^2) = 2, \qquad E_2 = E(xy) = \rho.$$

The invariant E_1 is useless, being simply a numerical constant. Thus we have only one relevant invariant $E_2 = \rho$ having the standard range $(-1, 1)$. $dE_2 = d\rho$, so that the rule gives uniform prior probability for the correlation coefficient.

Comparing the present invariance rule with the previous one given in Sec. 5.6, we note that the present rule leads directly to the appropriate prior-probability forms in the foregoing five cases. The previous rule leads to Haldane's form $d\alpha/\alpha(1-\alpha)$ for the binomial and the chance parameter, and then we have to combine it with J in order to have uniform prior probability for α.

8.4.6 Several parameters

For distributions with many parameters we do not get a single invariance rule that will apply satisfactorily to a large number of distributions. We can, however, express the appropriate prior-probability forms in terms of the invariants E_k, as we did previously in terms of the u_k. However, it is found that in the case of several parameters the previous system of invariants is simpler and more convenient than the present one. We shall therefore not take up the question of expressing the well-known prior-probability forms in terms of the E_k's.

Finally the alternative system of invariants can be extended to distributions of Sec. 7 also, where we may now select g(x) and take its expectation.

REFERENCES

Adke, S. R. (1958): A note on the distance between two populations, Sankhya, Ind. J. Statistics 19, 195.

Bayes, T. (1763): An essay towards solving a problem in the doctrine of chances, Phil. Trans. 53, 370.

Bhattacharya, A. (1942): On discrimination and divergence, Proc. Ind. Sci. Congr., Part III, 13.

Broad, C. D. (1918): The relation between induction and probability, Mind 27, 389.

Diananda, P. H. (1946): Unpublished, but mentioned by Jeffreys (1948, p. 169).

Edgeworth, F. Y. (1908): On the probable errors of frequency constants, J. Roy. Stat. Soc. 71, 651.

Elderton, W. P. (1938): Frequency Curves and Correlation, 3rd ed., London, Cambridge Univ. Press.

Fisher, R. A. (1922): On the mathematical foundations of theoretical statistics, Phil. Trans. A 222, 309.

_____ (1934): Two new properties of mathematical likelihood, Proc. Roy. Soc. A 144, 285.

_____ (1944): Statistical Methods for Research Workers, 9th ed., London, Oliver and Boyd.

Haldane, J. B. S. (1932): A note on inverse probability, Proc. Camb. Phil. Soc. 28, 55.

Huzurbazar, V. S. (1948): Inverse probability and sufficient statistics, Proc. Camb. Phil. Soc. 45, 225.

_____ (1949a): Some Properties of Distributions Admitting Sufficient Statistics, Ph.D. Thesis, Cambridge University (Part unpublished).

_____ (1949b): On a property of distributions admitting sufficient statistics, Biometrika 36, 71.

_____ (1955a): Confidence intervals for the parameter of a distribution admitting a sufficient statistics when the range depends on the parameter, J. Roy. Stat. Soc. B 17, 86.

_____ (1955b): Exact forms of some invariants for distributions admitting sufficient statistics, Biometrika 42, 533.

_____ (1955c): On the certainty of an inductive inference, Proc. Camb. Phil. Soc. 51, 761.

_____ (1955d): Some properties of the invariant I_1, J. Univer. Poona: Sci. Sec. 6, 115.

_____ (1956): Sufficient statistics and orthogonal parameters, Sankhya, Ind. J. Stat. 17, 217.

Jeffreys, H. (1946): An invariant form for the prior probability in estimation problems, Proc. Roy. Soc. A 186, 453.

_____ (1948): Theory of Probability, 2nd ed., Oxford, Clarendon Press.

_____ (1954): The present position in probability theory, Brit. J. Phil. Sci. 5, 275.

_____ (1957): Scientific Inference, 2nd ed., London, Cambridge Univ. Press.

Kendall, M. G. (1946): The Advanced Theory of Statistics, London, vol. 2, Charles Griffin & Co. Ltd.

Koopman, B. O. (1936): On distributions admitting a sufficient statistic, Trans. Amer. Math. Soc. 39, 399.

Kullback, S., and Leibler, R. A. (1951): On information and sufficiency, Ann. Math. Stat. 22, 79.

Mahalanobis, P. C. (1936): On the generalized distance in statistics, Proc. Natl. Inst. Sci. Ind. 12, 49.

Neyman, J. (1942): Basic ideas and some recent results of the theory of testing statistical hypotheses, J. Roy. Stat. Soc. 105, 292.

Pitman, E. J. G. (1936): Sufficient statistics and intrinsic accuracy, Proc. Camb. Phil. Soc. 32, 567.

Rao, B. R. (1959a): Contributions to the theory of statistical estimation, Ph.D. Thesis, Poona Univ. (Part unpublished).

_____ (1959b): Properties of the invariant I_m (m odd) for distributions admitting sufficient statistics. Sankhya 21, 355.

PART 2

THE GENERAL FORMS OF DISTRIBUTIONS ADMITTING SUFFICIENT STATISTICS FOR PARAMETERS IN NONREGULAR CASES

1. Introduction

1.1 General Introduction to Results Obtained

The main purpose of Part 2 of this book is to obtain rigorous proofs of the general forms of univariate distributions possessing probability density functions which admit sufficient statistics for parameters in nonregular cases (as defined in Sec. 2.9). The assumptions made in the derivation of the general forms have been explicitly stated. It is interesting to note that the general forms in nonregular cases with a single unknown parameter were surmised with remarkable intuition by Pitman (1936), and a criticism of his derivation will be found in Sec. 1.2.

Section 2 contains statements of definitions and derivation of some useful preliminary results. Section 2.4 is concerned with definition and properties of two inverse functions defined in relation to a continuous monotonic function. These inverse functions are found very useful in subsequent investigations. Section 2.5 is devoted to the general solution of a functional equation which occurs frequently in the determination of the general form of distributions. It should be noted that results of Sec. 2.5 are obtained without any continuity or differentiability restrictions of the function f. Section 2.6 is concerned with derivation of the joint sampling distribution of the smallest, the greatest, and any unordered member in random samples of size $n \geq 2$ from a distribution with p.d.f. (probability density function) $f(x)$. This is a basic distribution which immediately yields the conditional distributions leading to the general forms of distributions admitting sufficient statistics in some important nonregular cases.

Section 3 is occupied with investigation of the general form of the p.d.f. $f(x, \theta)$ of the family of distributions depending on a single unknown parameter θ which admits a single sufficient statistic for all samples of any size $n \geq 2$ when the ranges $a[(\theta), b(\theta)]$ of the family of distributions depend on θ. As remarked in Sec. 2.1, the case $n = 1$ is trivial in which every family of distributions depending on a single unknown parameter θ admits a single sufficient statistic for θ. Theorem 6 of Sec. 3.9 is the central theorem of this discussion, which gives the general form (3-145) and the single sufficient statistic defined by either Eq. (3-146) or Eq. (3-147). Incidentally the sampling distribution of the single sufficient statistic is obtained and is given by Eqs. (3-127) and (3-128) or by Eqs. (3-129) and (3-130). Pitman's conjecture about nonexistence of a single sufficient statistic when $a(\theta)$ and $b(\theta)$ are monotonic in θ but not in opposite senses is proved in Theorem 6 of Sec. 3.9.

The results of Sec. 3 have been obtained under fairly general conditions, mainly on account of the use of the inverse functions of Sec. 2.3 and results of Sec. 2.5. The terminals $a(\theta)$ and $b(\theta)$ need not be strictly monotonic functions of θ for all $\theta \varepsilon \Omega$, where Ω is the interval of admissible values of θ. The results obtained hold good even if max $a(\theta) \neq$ min $b(\theta)$ for all $\theta \varepsilon \Omega$, and also if any of the sample values do not fall within the range of values of either $a(\theta)$ or $b(\theta)$. No continuity or differentiability restrictions have been imposed on the p.d.f. of the family of distributions in results of Sec. 3.

Section 5 deals with the nonregular case with a single unknown parameter θ which does not admit a single sufficient statistic for θ, for $n \geq 2$ but which admits a minimal set of $\nu \geq 2$ jointly sufficient statistics for θ for all $n > \nu$. Section 5.1 deals with the case $\nu = 2$, and Sec. 5.2 deals with the case $\nu > 2$.

Section 7 is concerned with the nonregular case with two parameters (θ_1, θ_2) admitting a minimal set of $\nu \geq 2$ jointly sufficient statistics for (θ_1, θ_2) for all $n > \nu$. Section 7.1 deals with the case $\nu = 2$, and Sec. 7.2 deals with $\nu > 2$.

Section 8 deals with the nonregular case involving $p > 2$ parameters.

The nonregular cases considered fall into two categories of the "pure" and "mixed" types (as defined in Sec. 2.9). Cases discussed in Sec. 3, Sec. 5.1 (Case II), and Sec. 7.1 (Case I) belong to the pure type, whereas cases discussed in Sec. 5.1 (Case I),

Sec. 5.2, Sec. 7.1 (Case II), Sec. 7.2, and Sec. 8 belong to the mixed type. The general forms of distributions in pure-type cases have been derived from the Form A equivalents of the basic definitions of sufficient statistics in Sec. 2.1. As remarked in Sec. 2.1, these forms are equivalent to the corresponding basic definitions, without any restrictions. In particular, it may be emphasized that the results obtained in the pure-type cases are independent of the use of the Factorability Criterion (Form B) which is equivalent to the basic definitions of sufficient statistics under certain restrictions only (See Sec. 2.1). The results obtained in the pure-type cases are free from continuity or differentiability restrictions on the p.d.f. of the family of distributions.

In mixed-type cases, one or two members of the minimal set of sufficient statistics are defined in terms of only the smallest and the greatest members in the sample. In search of the remaining members of the minimal set of sufficient statistics the general forms (2-3) and (2-4) for the regular case are assumed as necessary conditions and then the general forms are deduced for the mixed-type cases. Since the general forms (2-3) and (2-4) of the regular cases involve the assumptions of Koopman regularity conditions (as mentioned in Sec. 2.1), which include in particular existence of certain partial derivatives of the second order of the p.d.f. of the family of distributions, as well as the applicability of the Factorability Criterion, the results obtained in mixed-type cases are under the assumptions of the Koopman regularity conditions. It will be realized then that an interesting feature of this discussion is the treatment of the pure nonregular cases on their own merits, rather than making them dependent on the regular cases by imposition of additional Koopman regularity conditions.

However, it is also shown in Sec. 4, Sec. 6 (Case II) and Sec. 7.1.1 (concluding remarks) that if in addition the Koopman regularity conditions are assumed for the pure-type cases, the corresponding general forms can be obtained with considerable ease.

As regards the mixed-type cases, each case is first treated with minimum use of the Koopman regularity conditions, i.e., by assuming the Koopman regularity conditions only with regard to those members of the minimal set of sufficient statistics which are not provided by the smallest and the greatest members in the sample. Then in Secs. 5, 7, and 8 (concluding remarks), somewhat simpler alternative derivations of the general forms in mixed-type cases are given or suggested, assuming in full the Koopman regularity conditions with regard to all members of the minimal set of sufficient statistics.

1.2 Review of Previous Work

Although the general forms of the p.d.f. in nonregular cases with a single unknown parameter derived in this work were given by Pitman (1936), there are inaccuracies, incompleteness, and conjecture in Pitman's derivation of the general forms in question. Moreover, the assumptions have not been clearly stated by him. Pitman first assumes that the upper terminal b of the range is fixed and the lower terminal a is a function of θ. If T denotes the smallest member in a sample, then a sufficient statistic when it exists must be T (or a function of T). Pitman argues: "When T is fixed, the probability that any particular member of the sample, say x_1 (which we may think of as the first member drawn), takes the value $x \geq T$ is still proportional fo $f(x; \theta)$. Since this restricted distribution of x_1 is independent of θ, $f(x_1, \theta)/f(x_1', \theta)$ must be independent of θ. Therefore $f(x, \theta) = g(x)/h(\theta)$."

Thus Pitman has not actually derived the conditional distribution of x_1 for a given T, but he has guessed its form by intuition. This intuitive argument is also not quite accurate since, as will be seen as a particular case of the general conditional distribution of x_i for a given T obtained in Sec. 3.6, the conditional distribution of x_1 for a fixed T is given by Eqs. (1-1) and (1-2):

$$P(x_1; \theta | T) = \frac{n-1}{n} \frac{f(x_1; \theta)}{\int_{a(T)}^{b} f(x; \theta)\,dx}, \qquad [a(T) < x_1 \le b], \tag{1-1}$$

and $$P[x_1 = a(T); \theta | T] = \frac{1}{n}, \tag{1-2}$$

i.e., the conditional distribution of x_1 for a fixed T is of a mixed type with p.d.f. given by Eq. (1-1) in the half-open interval $[a(T) < x_1 \le b]$ and with the probability given by Eq. (1-2) at the discrete mass point $x_1 = a(T)$. These facts and also the fact that Eq. (1-2) is independent of θ are a serious omission in Pitman's proof for the particular case considered by him. Further, for the general case when both the terminals a and b of the range depend on θ, Pitman does not give a proof, but takes $a(\theta) = \theta$ and assumes that $b(\theta)$ is a strictly decreasing function of θ. Let $x_{(1)}$ and $x_{(n)}$ be respectively the smallest and the greatest members in the sample. Pitman argues that a sufficient statistic, if it exists, must be T (or a function of T), where T is the smaller of $x_{(1)}$ and $b^{-1}[x_{(n)}]$, where b^{-1} is the inverse of the function $b(\theta)$. He remarks: "Samples with a fixed T lie in a fixed range, and T is a sufficient statistic provided that the frequency function of the population is of the form $f(x)/F(\theta)$, $\theta \le x \le b(\theta)$. It would seem that this is the only type of case where a sufficient statistic exists." It is thus clear that Pitman assumes without proof the general form $f(x)/F(\theta)$ of the p.d.f. in this general case and also concludes without proof that T is a sufficient statistic. Even if a proof is supplied of the statement that T is a sufficient statistic, assuming that the p.d.f. is of the form $f(x)/F(\theta)$, it will ensure only sufficiency of that form, and the question of necessity of the form will still remain open. As will be seen from the proofs we shall give, the proof of the necessity of the general form is more difficult than that of the sufficiency of the general form. The last sentence in the quotation from Pitman is again of the nature of a conjecture, which we have restated and proved in Theorem 2 of Sec. 3.9. For the nonregular cases treated in Sec. 5.1, Pitman has given the general forms without rigorous proofs.

Davis (1951) has obtained the form (3-145) for the nonregular case with a single sufficient statistic for a single unknown parameter, and also has given a proof of Pitman's conjecture. Though Davis has stated his assumptions carefully, the following main objections can be raised against his work:

1. Davis has throughout in his paper assumed the general form (2-3) for the regular case with a single sufficient statistic for a single unknown parameter, and then deduced the form (3-145) for the nonregular case. The results obtained by him are thus subject to Koopman regularity conditions.

2. A more serious objection is that Davis takes the definition of a single sufficient statistic in the following form: "A statistic T is sufficient for a parameter θ if, and only if, the conditional distribution of the *ordered* sample values $x_{(1)}, \ldots, x_{(n)}$ for a fixed T is independent of θ." Since, in the basic definition of a sufficient statistic (Sec. 2.1), or its equivalent Forms A and B, the sample values $x_1, \ldots, x_n$ need not be ordered, the argument used by Davis may at best be regarded as leading to the necessity of the form (3-145), but not toward sufficiency of that form.

3. When only one terminal, say a, of the range (a, b) depends on θ, Davis limits the scope of his investigation only to the case when the sufficient statistic is $x_{(1)}$. Further, when both a and b depend on θ, he limits the scope of his investigation only to the case when the single sufficient statistic is a function of both $x_{(1)}$ and $x_{(n)}$. He does not consider other possibilities. In particular, the "range part" of the basic definition of a sufficient statistic, (Sec. 2.1), and its equivalent Forms A and B, specially emphasized by us in the statements, seems to be absent in his investigation. As will be seen from the proofs we shall give, in the primary search for a sufficient statistic in nonregular cases, it is this "range part" of the definition of a sufficient statistic which acts as a committee with the power of nominating some statistic as a candidate for subsequent election to the office of a sufficient statistic.

When the range depends on a single parameter θ, both Pitman and Davis have assumed that each terminal (when it is not independent of θ) is a *strictly* monotonic function of θ. Consequently both of them have treated separately both cases (1) when both terminals depend on θ and (2) when only one terminal depends on θ. The results we shall present allow the possibility of any of the two terminals remaining constant in subintervals of asmissible values of θ, and this has also enabled us to deduce the case (2) in point as a particular case of case (1) in point.

It has been assumed explicitly by Pitman, and tacitly by Davis, the condition

$$\max a(\theta) = \min b(\theta) \tag{1-3}$$

for all admissible values of θ. Here it needs to be pointed out that the inverse functions $a^{-1}[x_{(1)}]$ and $b^{-1}[x_{(n)}]$ used by Pitman and Davis in the definition of a sufficient statistic are not defined if either $x_{(1)} > \max a(\theta)$ or $x_{(n)} < \min b(\theta)$, even if the condition (1-3) is assumed. The results we shall give are free from the restriction (1-3), and moreover introduction of two inverse functions (Sec. 2.3) defined for *all* real values of their arguments has enabled us to obtain results of a very general character, and the definition of the single sufficient statistic is given by the *same* formula Eq.[(3-146)] for the monotonic Type I case, and the formula (3-147) for the monotonic Type II case, whether $x_{(1)}$ falls within or outside the range of values of $a(\theta)$, and $x_{(n)}$ falls within or outside the range of values of $b(\theta)$.

In connection with the assumption (2) of Sec. 3.1 that $a(\theta)$ and $b(\theta)$ are continuous and differentiable functions of θ for all $\theta \in \Omega$, it may be remarked that this assumption is also implicit in the papers by Pitman and Davis. For in the general case, both have assumed $a(\theta) = \theta$ "for convenience", which implies the assumption that $a(\theta)$ is a strictly increasing, continuous, and differentiable function of θ for all θ. Further, as pointed out in Sec. 3.1, the assumption that one of the terminals $a(\theta)$ or $b(\theta)$ is a continuous and differentiable function of θ implies that the other terminal is also a continuous and differentiable function of θ almost everywhere in Ω.

Finally it may be noted that the alternative derivations of the general forms in nonregular cases with the additional assumptions of the Koopman regularity conditions, given in the present work, are much simpler in comparison with the derivation of Davis for the single-parameter case.

2. Definitions, Notation, and Preliminary Results

2.1 Sufficient Statistics

For the purposes of this discussion, we shall adopt the following definitions.

Definition of a statistic

If $x_1, \ldots, x_n$ is a random sample of n independent members from a distribution depending on some unknown parameters, any single-valued function of $x_1, \ldots, x_n$ (not depending on any of the unknown parameters) is called a statistic.

Basic definition of a single sufficient statistic for a single unknown parameter

A statistic T is said to be a single sufficient statistic for a single unknown parameter θ if the conditional distribution of any other statistic T_1 for every fixed T is independent of θ, both in form and range, for all $\theta \in \Omega$, where Ω is the set of admissible values of θ.

The definition also includes the nonregular case when the range of the distribution depends on θ, since the definition emphasizes that the range of the conditional distribution of T_1 for every fixed T is independent of θ, for all $\theta \in \Omega$.

In the application of the definition of a single sufficient statistic for a single parameter θ, we shall assume that n (sample size) ≥ 2. The case $n = 1$ is trivial wherein any two statistics are functionally related, and hence the conditional distribution of one statistic when the other statistic is fixed is ipso facto independent of θ. Thus, for samples of size $n = 1$, every family of distributions depending on a single parameter θ admits a single sufficient statistic for θ. Consequently, in our investigation of the general form of distributions admitting a single sufficient statistic for a single parameter θ, it will be assumed that $n \geq 2$.

From the basic definition of a single sufficient statistic, several other useful equivalent forms (some of them requiring certain restrictions) have been derived, two of which will now be given. In Forms A and B, which follow, $x_1, \ldots, x_n$ is a random sample of n members from a distribution depending on a single parameter θ.

Form A

A statistic T is a single sufficient statistic for a single unknown parameter θ if the conditional distribution of each x_i $(i = 1, \ldots, n)$ for every fixed T is independent of θ, both in form and range, for all $\theta \in \Omega$.

It is interesting to note that, though at first sight Form A may appear to be a particular case of the basic definition where T_1 is taken to be x_i itself, it can be easily shown that if the conditional distribution of each x_i $(i = 1, \ldots, n)$ for a fixed T is independent of θ in both form and range, then the conditional distribution of any other statistic T_1 for a fixed T is also independent of θ in both form and range. Hence Form A is equivalent (without any restrictions) to the basic definition. We shall find Form A especially useful.

Form B: The Factorability Criterion (or Neyman Criterion)

If $L(x_1, \ldots, x_n; \theta)$ denotes the likelihood function of the sample, T is a single sufficient statistic for θ if the range of each x_i $(i = 1, \ldots, n)$ for every fixed T is independent of θ and if L is expressible in the form

$$L(x_1, \ldots, x_n; \theta) = P(T, \theta) \quad \phi(x_1, \ldots, x_n) \tag{2-1}$$

for all $x_1, \ldots, x_n$ belonging to R and for all $\theta \in \Omega$, where $P(T, \theta)$ is the p.d.f. of T, and $\phi(x_1, \ldots, x_n)$ is independent of θ.

The Factorability Criterion itself is often stated in another equivalent form,

$$L(x_1, \ldots, x_n; \theta) = \phi_1(T, \theta) \quad \phi_2(x_1, \ldots, x_n), \tag{2-2}$$

where $\phi_1(T, \theta)$ need not denote the p.d.f. of T.

The equivalence between the basic definition and the Factorability Criterion was proved under certain conditions by Neyman (1935), and under much more general conditions by Halmos and Savage (1949).

It is unfortunate that the condition that "the range of each x_i $(i = 1, \ldots, n)$ for a fixed T is independent of θ" is often not explicitly mentioned in the statements of the Factorability Criterion commonly found in the literature, which assume tacitly that the range of the distribution does not depend on θ. If explicit mention is made of this condition in the statement of the Factorability Criterion, it covers the nonregular case also.

2.2 Basic Definitions of a Set of Sufficient Statistics for a Set of Unknown Parameters

A set $(T_1, \ldots, T_q)$ of q statistics is said to be sufficient (or jointly sufficient) for a set $(\theta_1, \ldots, \theta_p)$ of p unknown parameters if the conditional distribution of any other statistic T for given $(T_1, \ldots, T_q)$ is independent of $(\theta_1, \ldots, \theta_p)$ in both form and range for all $(\theta_1, \ldots, \theta_p) \in \Omega$.

A set of q sufficient statistics for a set of p unknown parameters is said to be a minimal set of sufficient statistics if there exists no set of sufficient statistics whose number is smaller than q.

Two particular cases of interest are: (1) $p = 1, q = 2$: a minimal pair of sufficient statistics for a single unknown parameter; and (2) $p = 2, q = 2$: a minimal pair of sufficient statistics for a pair of unknown parameters.

Equivalent Forms A and B of the above definitions analogous to the equivalent Forms A and B of the basic definition of a single statistic for a single unknown parameter follow on similar lines.

In our investigation of the general forms of distributions admitting a minimal set of q sufficient statistics, we shall assume (to avoid triviality) that $n > q$, since the n sample values constitute a set of sufficient statistics for the parameters of every distribution.

2.3 Statements of the General Forms of Distributions Admitting Sufficient Statistics for Parameters in Regular Cases

Case I: The Regular Case with a Single Sufficient Statistic for a Single Unknown Parameter

Let $f(x; \theta)$ be the p.d.f. of a family of distributions depending on a single unknown parameter θ whose admissible values belong to some nondegenerate interval Ω. Let the range (a, b) of the distribution of x be independent of θ. Under certain conditions, the general form of $f(x; \theta)$ admitting a single sufficient statistic for θ for all $\theta \in \Omega$ and for all random samples of any size $n \geq 2$ is

$$f(x; \theta) = \exp[u(\theta)\, v(x) + A(x) + B(\theta)] \quad (a \leq x \leq b), \qquad (2\text{-}3)$$

where u and B are functions of θ only, and v and A are functions of x only. The form (2-3) was given by Koopman (1936), and Darmois (1935). For a statement of the set of conditions under which the form (2-3) is obtained, see Koopman (1936). We shall use the phrase "the Koopman regularity conditions" for conditions under which the form (2-3) is derived. The Koopman regularity conditions assume in particular (1) the existence of the partial derivative $\partial^2 f/\partial\theta\, \partial x$ of the second order, and (2) the equivalence of the Factorability Criterion and the basic definition of a single sufficient statistic. Condition (1) apparently restricts the scope of distributions where both the variate x and the parameter θ are continuous. That this condition is unnecessary can be verified from the fact that all the well-known discrete distributions such as binomial and Poisson, which admit sufficient statistics for their parameters, have their probability functions of the form (2-3). Condition (1) is used in the necessity proof of the form (2-3), but it is not required for the sufficiency proof.

It may be remarked here that in the derivation of the form (2-3), the Factorability Criterion (Form B of the basic definition of a single sufficient statistic for a single unknown parameter is used, but the part of the criterion that the range of each x_i for a fixed T (T being a sufficient statistic) is independent of θ is not explicitly used in the derivation (i.e., the necessity part of the proof) of Eq. (2-3); however, this condition is implicitly used in the sufficiency proof of the form (2-3). For if $f(x; \theta)$ has the form (2-3), and if the range of the distribution of x is independent of θ, then $T = \sum_{i=1}^{n} v(x_i)$ is a sufficient statistic for θ, since it is at once verified that T satisfies the Factorability Criterion in full (i.e., in both range and form). On the other hand, if $f(x; \theta)$ is of form (2-3), but the range (a, b) depends on θ, then under the Koopman regularity conditions, $t = \sum_{i=1}^{n} v(x_i)$ satisfies the part of the Factorability Criterion relating to form, but may not necessarily satisfy the other part relating to the range. Hence, under the Koopman regularity conditions, the form (2-3) still remains a necessary condition for the existence of a single sufficient statistic for θ for all random samples of any size $n \geq 2$ and for all $\theta \in \Omega$, though it ceases to be a sufficient condition for the same. Moreover, it also follows from the necessity proof of form (2-3) that under the Koopman regularity conditions, even in the nonregular case, if a single sufficient statistic exists for θ for all $\theta \in \Omega$ and for all random samples of any size $n \geq 2$, it must be T_1 (or any one-to-one function of T_1), where $T_1 = \sum_{i=1}^{n} v(x_i)$.

Case II: The Regular Case with Several Jointly Sufficient Statistics and Several Unknown Parameters

In the regular case, if a family of distributions with p.d.f. $f(x; \theta_1, \ldots, \theta_p)$ depends on p unknown parameters $(\theta_1, \ldots, \theta_p)$, where the admissible values of $(\theta_1, \ldots, \theta_p)$ belong to some p-dimensional nondegenerate interval Ω, then under the Koopman regularity conditions the general form of $f(x; \theta_1, \ldots, \theta_p)$ admitting a minimal set of ν jointly sufficient statistics for $(\theta_1, \ldots, \theta_p)$ for all $(\theta_1, \ldots, \theta_p) \in \Omega$ and for all random samples of any size $n > \nu$ is

$$f(x; \theta_1, \ldots, \theta_p) = \exp\left[\sum_{k=1}^{\nu} u_k(\theta_1, \ldots, \theta_p)\, v_k(x) + A(x) + B(\theta_1, \ldots, \theta_p)\right] \qquad (a \leq x \leq b), \tag{2-4}$$

where the u_k $(k = 1, \ldots, \nu)$ are linearly independent functions of $(\theta_1, \ldots, \theta_p)$ only, B is a function of $(\theta_1, \ldots, \theta_p)$ only, the v_k $(k = 1, \ldots, \nu)$ are linearly independent functions of x only, and A is a function of x only. When the form (2-4) holds good, the set $T_k = \sum_{i=1}^{n} v_k(x_i)$ $(k = 1, \ldots, \nu)$ of statistics is minimal sufficient for the parameters $(\theta_1, \ldots, \theta_p)$. In the form (2-4) ν may be greater than, equal to, or less than p. The Koopman regularity conditions in this case assume in particular the existence of the second-order partial derivatives $\partial^2 f/\partial\theta_j\, \partial x$ $(j = 1, \ldots, p)$. As in Case I, the form (2-4) will be a necessary (though not sufficient) condition, even in the nonregular case, for existence of a minimal set of ν jointly sufficient statistics for $(\theta_1, \ldots, \theta_p)$, under the Koopman regularity conditions, for all $(\theta_1, \ldots, \theta_p) \in \Omega$ for all random samples of any size $n > \nu$; and the minimal set of sufficient statistics, when it exists, must be the $T_k = \sum_{i=1}^{n} v_k(x_i)$ $(k = 1, \ldots, \nu)$, or any set of one-to-one functions of the T_k.

2.4 Two Inverse Functions Associated with a Continuous Monotonic Function

Let $a(\theta)$ be a continuous, monotonic function of θ defined in some nondegenerate interval $\Omega = (\lambda_1, \lambda_2)$, which may be closed, open, halfopen, finite, or infinite. If Ω is open at λ_1, we shall use the symbol $a(\lambda_1)$ to denote $a(\lambda_1 + 0) = \lim_{\theta \to \lambda_1 + 0} a(\theta)$, which exists (either finite or infinite). Similarly, if Ω is open at λ_2, $a(\lambda_2 - 0) = \lim_{\theta \to \lambda_2 - 0} a(\theta)$. With these meanings for $a(\lambda_1)$ and $a(\lambda_2)$, it is easily seen that $a(\theta)$ will continue to be monotonic in the "extended closed interval" $\Omega = (\lambda_1, \lambda_2)$ if Ω is supposed to be extended by addition of its open endpoints, if any. In cases where Ω is so extended, the phrase "extended closed interval" will be invariably used in the sequel.

The interval of values of $a(\theta)$ is $I = [a(\lambda_1), a(\lambda_2)]$ or $[a(\lambda_2), a(\lambda_1)]$ according as $a(\theta)$ is m.i. (monotone increasing) or m.d. (monotone decreasing). The interval I may be closed, open, half-open, finite, or infinite.

For every $\alpha \in I$, the equation

$$a(\theta) = \alpha \tag{2-5}$$

has either a unique solution or there exists a nondegenerate closed interval δ_α of values of θ such that every point of δ_α is a solution of Eq. (2-5) according as $a(\theta)$ is strictly monotonic at the point of the unique solution or $a(\theta)$ = constant = α at every point of δ_α.

We shall use the symbols $a_1^{-1}(\alpha)$ and $a_2^{-1}(\alpha)$ to denote respectively the smallest and the greatest roots of Eq. (2-5). At every point α at which $a_1^{-1}(\alpha) = a_2^{-1}(\alpha)$, eq. (2-5) has a unique solution. The set of points α at which $a_1^{-1}(\alpha) \neq a_2^{-1}(\alpha)$ is of measure zero. Each of the functions $a_1^{-1}(\alpha)$ and $a_2^{-1}(\alpha)$ is a single-valued, *strictly* monotonic function of α defined in the interval I. The functions $a_1^{-1}(\alpha)$ and $a_2^{-1}(\alpha)$ may, however, have discontinuity points in sets of measure zero. For every $\alpha \in I$, the functions $a_1^{-1}(\alpha)$ and $a_2^{-1}(\alpha)$ possess the property

$$a[a_1^{-1}(\alpha)] = a[a_2^{-1}(\alpha)] = \alpha. \tag{2-6}$$

In the particular case when $a(\theta)$ = constant = k for all $\theta \in \Omega$, $a_1^{-1}(\alpha)$ and $a_2^{-1}(\alpha)$ are defined only for the single point $\alpha = k$ of the degenerate interval I, and

$$a_1^{-1}(k) = \lambda_1, \quad a_2^{-1}(k) = \lambda_2. \tag{2-7}$$

If $a(\theta)$ is strictly monotonic for all $\theta \in \Omega$, we may write

$$a_1^{-1}(\alpha) = a_2^{-1}(\alpha) = a^{-1}(\alpha) \quad \text{for all } \alpha \in I. \tag{2-8}$$

Now suppose that $a(\theta)$ is m.i. in Ω. Then for every $\alpha \in I$, the inequality

$$a(\theta) \leq \alpha \tag{2-9}$$

is equivalent to the inequality on θ,

$$\theta \leq a_2^{-1}(\alpha). \tag{2-10}$$

The inequality

$$a(\theta) \geq \alpha \tag{2-11}$$

is equivalent to the inequality

$$\theta \geq a_1^{-1}(\alpha). \tag{2-12}$$

Similarly, if $a(\theta)$ is m.d.,

$$a(\theta) \leq \alpha \quad \text{is equivalent to} \quad \theta \geq a_1^{-1}(\alpha), \tag{2-13}$$

and $$a(\theta) \geq \alpha \quad \text{is equivalent to} \quad \theta \leq a_2^{-1}(\alpha). \tag{2-14}$$

When I is a proper subset of the interval $R_1 = (-\infty, \infty)$, it will be found convenient in our work to extend the definitions of $a_1^{-1}(\alpha)$ and $a_2^{-1}(\alpha)$, where α is any real number not belonging to I. Let C(I) denote the complement of I with respect to R_1. For every $\alpha \in C(I)$, the equation $a(\theta) = \alpha$ has no real solution for θ in Ω, and we define

$$a_1^{-1}(\alpha) = \lambda_1, \quad a_2^{-1}(\alpha) = \lambda_2 \qquad [\alpha \in C(I)]. \tag{2-15}$$

Hence $a_1^{-1}(\alpha)$ and $a_2^{-1}(\alpha)$ assume constant values λ_1 and λ_2 respectively in C(I). The functions $a_1^{-1}(\alpha)$ and $a_2^{-1}(\alpha)$ are thus defined for all α in R_1, and they are single-valued. Moreover, the interval of values assumed by each of $a_1^{-1}(\alpha)$ and $a_2^{-1}(\alpha)$ in R_1 is the *extended closed interval* $\Omega = (\lambda_1, \lambda_2)$.

We note from Eq. (2-15) that

$$a[a_1^{-1}(\alpha)] = a(\lambda_1) \qquad [\alpha \in C(I)], \tag{2-16}$$

and $$a[a_2^{-1}(\alpha)] = a(\lambda_2) \qquad [\alpha \in C(I)]. \tag{2-17}$$

Equations (2-16) and (2-17) show that Eq. (2-6) does not hold good if $\alpha \in C(I)$. We also note that

$$a_1^{-1}[a(\lambda_1)] = \lambda_1, \tag{2-18}$$

and $$a_2^{-1}[a(\lambda_2)] = \lambda_2. \tag{2-19}$$

Equation (2-18) implies: Whether the borderpoint $a(\lambda_1)$ belongs to I or not, a_1^{-1} has the same value λ_1 there. Similarly Eq. (2-19) implies: whether the border point $a(\lambda_2)$ belongs to I or not, a_2^{-1} has the same value λ_2 there.

To prove Eq. (2-18) it follows at once from the definition (2-15) if $a(\lambda_1) \in C(I)$. If $a(\lambda_1) \in I$, λ_1 is the smallest root of the equation $a(\theta) = a(\lambda_1)$, so that Eq. (2-18) holds good. Similarly Eq. (2-19) is proved.

If $a(\theta)$ is m.i., the inequality

$$a(\theta) \leq \alpha \qquad [\alpha \geq a(\lambda_2)] \tag{2-20}$$

is satisfied for all θ in Ω, i.e., for all $\theta \leq \lambda_2$, i.e., for all

$$\theta \leq a_2^{-1}(\alpha) \qquad [\alpha \geq a(\lambda_2)], \tag{2-21}$$

since, if $\alpha > a(\lambda_2)$, $\alpha \in C(I)$ and so $a_2^{-1}(\alpha) = \lambda_2$; and if $\alpha = a(\lambda_2)$, $a_2^{-1}[a(\lambda_2)] = \lambda_2$ in virtue of Eq. (2-19). Hence Eqs. (2-20) and (2-21) are equivalent statements. Similarly it is easily seen that the inequality

$$a(\theta) \geq \alpha \qquad [\alpha \leq a(\lambda_1)] \tag{2-22}$$

is equivalent to the inequality

$$\theta \geq a_1^{-1}(\alpha) \qquad [\alpha \leq a(\lambda_1)]. \tag{2-23}$$

If $a(\theta)$ is m.d.,

$$a(\theta) \leq \alpha \qquad [\alpha \geq a(\lambda_1)] \tag{2-24}$$

is equivalent to

$$\theta \geq a_1^{-1}(\alpha) \qquad [\alpha \geq a(\lambda_1)]; \tag{2-25}$$

and $a(\theta) \geq \alpha \quad [\alpha \leq a(\lambda_2)]$ (2-26)

is equivalent to

$$\theta \leq a_2^{-1}(\alpha) \quad [\alpha \leq a(\lambda_2)]. \tag{2-27}$$

Thus the four equivalence relations for inequalities, proved previously for all α in I, are extended to all α in C(I), and hence they hold good for all α in R_1.

Incidentally, it may be noted that our extended definitions of $a_1^{-1}(\alpha)$ and $a_2^{-1}(\alpha)$ may also be regarded as being respectively the lower and the upper limits of the "roots" of either of the inequalities

$$a(\theta) \geq \alpha, \quad a(\theta) \leq \alpha \quad [\alpha \in C(I)]. \tag{2-28}$$

Since every point of $\Omega = (\lambda_1, \lambda_2)$ is a solution of either inequality in Eq. (2-28), λ_1 and λ_2 are respectively the lower and the upper limits of the roots of either inequality. Thus, in the extended definitions of $a_1^{-1}(\alpha)$ and $a_2^{-1}(\alpha)$, Eq. (2-5) is replaced by inequalities (2-28).

2.5 A Functional Equation

Let f(x, y) be a function of two independent variables x and y defined over a two-dimensional set S of values of (x, y). Let S_1 and S_2 be respectively one-dimensional sets of values of x and y, where (x, y) belongs to S. We prove that the general solution of f(x, y) satisfying the relation

$$\frac{f(x_1, y)}{f(x_2, y)} \quad \text{is independent of y, for all } (x_1, y), \text{ and } (x_2, y) \text{ belonging to S} \tag{2-29}$$

is as follows: For all (x, y) belonging to S, f(x, y) is of the form

$$f(x, y) = g(x)\, h(y), \tag{2-30}$$

where g(x) is a function of x defined over S_1, and h(y) is a function of y defined over S_2.

Proof. First suppose that $f(x, y)$ satisfies the relation (2-29) so that

$$\frac{f(x_1, y)}{f(x_2, y)} = \phi(x_1, x_2),$$

where $\phi(x_1, x_2)$ is independent of y. Then

$$f(x_1, y) = \phi(x_1, x_2)\, f(x_2, y). \tag{2-31}$$

Since Eq. (2-31) holds good for all (x_1, y) and (x_2, y) belonging to S, put $x_1 = x$ and keep x_2 fixed. Then for all $(x, y) \in S$ we have

$$\begin{aligned} f(x, y) &= \phi(x, x_2)\, f(x_2, y) \\ &= g(x)\, h(y), \end{aligned}$$

where $g(x) = \phi(x, x_2)$ is a function of x defined over S_1, and $h(y) = f(x_2, y)$ is a function of y defined over S_2. Thus $f(x, y)$ satisfies Eq. (2-30).

Next let $f(x, y)$ satisfy Eq. (2-30) for all $(x, y) \in S$, where $x \in S_1$ and $y \in S_2$. Then for all $(x_1, y) \in S$ and $(x_2, y) \in S$ we have

$$\frac{f(x_1, y)}{f(x_2, y)} = \frac{g(x_1)\, h(y)}{g(x_2)\, h(y)} = \frac{g(x_1)}{g(x_2)},$$

which is independent of y. Hence $f(x, y)$ satisfies the relation (2-29).

Corollary. Let $f(x, y_1, y_2, \ldots, y_n)$ be a function of $n + 1$ independent variables $x, y_1, y_2, \ldots, y_n$ defined over an $(n + 1)$-dimensional set S of values of $(x, y_1, y_2, \ldots, y_n)$. Let S_1 and S_2 be respectively one-dimensional and n-dimensional sets of values of x and $(y_1, \ldots, y_n)$, where $(x, y_1, \ldots, y_n)$ belongs to S. Then the general solution of $f(x, y_1, \ldots, y_n)$ satisfying the relation

$$\frac{f(x_1, y_1, \ldots, y_n)}{f(x_2, y_1, \ldots, y_n)} \quad \text{is independent of } (y_1, \ldots, y_n) \text{ for all } (x_1, y_1, \ldots, y_n) \text{ and } (x_2, y_1, \ldots, y_n) \text{ belonging to } S \tag{2-32}$$

is as follows: For all $(x, y_1, y_2, \ldots, y_n)$ belonging to S, $f(x, y_1, \ldots, y_n)$ is of the form

$$f(x, y_1, \ldots, y_n) = g(x)\ h(y_1, \ldots, y_n), \tag{2-33}$$

where g(x) is a function of x defined over S_1, and $h(y_1, \ldots, y_n)$ is a function of $(y_1, \ldots, y_n)$ defined over S_2.

The proof is exactly on similar lines as the proof of Eq. (2-29) already considered.

2.6 The Joint Sampling Distribution of the Smallest, the Greatest, and Any Unordered Member in Random Samples of Size $n \geq 2$ From a Distribution with p.d.f. f(x)

Let $x_{(1)}$ and $x_{(n)}$ be respectively the smallest and the greatest members in random samples of size $n \geq 2$ from a distribution with p.d.f. f(x). Let x_i be the ith unordered member in the sample. We regard x_i as the ith member drawn in a random sample of n independent members. Having chosen beforehand any particular value for i from the set (1, 2, ..., n) of values, we keep the chosen value of i fixed throughout. We seek to obtain the p.d.f. of the joint sampling distribution of $[x_{(1)}, x_{(n)}, x_i]$. It should be noted that $[x_{(1)}, x_{(n)}, x_i]$ is a mixture of two ordered statistics $x_{(1)}$ and $x_{(n)}$, and one unordered sample member x_i.

The unordered sample members $x_1, \ldots, x_n$ are a set of independent random variables, each having the same distribution as the parent distribution itself. On account of the symmetry inherent in the set of unordered sample members, the joint sampling distribution of $[x_{(1)}, x_{(n)}, x_i]$ and other sampling distributions involving x_i which we shall derive later will remain unaltered if x_i is replaced by x_j $(j = 1, \ldots, n, j \neq i)$.

First we shall make two cases for derivation of the joint sampling distribution of $[x_{(1)}, x_{(n)}, x_i]$: Case I, where $n > 2$; and Case II, where $n = 2$. We first consider Case I.

CASE I: $n > 2$.

It will be necessary to divide the region of possible values of $[x_{(1)}, x_{(n)}, x_i]$ into four mutually exclusive regions in the three-dimensional space of $[x_{(1)}, x_{(n)}, x_i]$:

Region 1: $x_{(1)} < x_i < x_{(n)}$,

Region 2: $x_{(1)} = x_i < x_{(n)}$,

Region 3: $x_{(1)} < x_i = x_{(n)}$,

Region 4: $x_{(1)} = x_i = x_{(n)}$.

Since the parent distribution has a p.d.f. $f(x)$,

$$P[x_{(1)} = x_i = x_{(n)}] = [x_1 = x_2 = \dots = x_i = \dots = x_n] = 0.$$

Hence the probability measure of Region 4 is zero.

The probability measure of Region 2 is $1/n$, since $P[x_{(1)} = x_i < x_{(n)}]$ is the same for all $i = 1, \dots, n$; at least one of the x_i must be $x_{(1)}$, and moreover the probability that two or more of the x_i coincide is zero. Similarly the probability measure of Region 3 is also $1/n$. Hence the probability measure of Region 1 is $(n - 2)/n$.

The joint sampling distribution of $[x_{(1)}, x_{(n)}, x_i]$ takes different forms in the different regions, which we shall now consider.

REGION 1: $x_{(1)} < x_i < x_{(n)}$.

In this three-dimensional region of probability measure $(n - 2)/n$, the three-dimensional joint distribution of $[x_{(1)}, x_{(n)}, x_i]$ is nondegenerate, and we shall use the symbol $P[x_{(1)}, x_{(n)}, x_i]$ to denote its p.d.f. The probability differential of this joint distribution is $P[x_{(1)}, x_{(n)}, x_i]\, dx_{(1)}\, dx_{(n)}\, dx_i$.

Let r be the number of sample members in the open interval $[x_{(1)}, x_i]$ and s be the number of sample members in the open interval $[x_i, x_{(n)}]$. Then $r + s = n - 3 \geq 0$, since we have assumed $n \geq 3$ in the case in point. r is a random variable which assumes the values 0, 1, ..., n - 3. For a given r, the probability that x_i falls in the range $(x_i - \frac{1}{2}\,dx_i,\ x_i + \frac{1}{2}\,dx_i)$, and of the remaining (n - 1) members in the random sample, just one member falls in the range $[x_{(1)} - \frac{1}{2}\,dx_{(1)},\ x_{(1)} + \frac{1}{2}\,dx_{(1)}]$, just r members fall in the range $[x_{(1)} + \frac{1}{2}\,dx_{(1)},\ x_i - \frac{1}{2}\,dx_i]$, just s members fall in the range $[x_i + \frac{1}{2}\,dx_i,\ x_{(n)} - \frac{1}{2}\,dx_{(n)}]$, and just one member falls in the range $[x_{(n)} - \frac{1}{2}\,dx_{(n)},\ x_{(n)} + \frac{1}{2}\,dx_{(n)}]$ is, to order $dx_{(1)}\ dx_{(n)}\ dx_i$,

$$p_r = \frac{(n-1)!}{1!r!s!1!}\left[\int_{x_{(1)}}^{x_i} f(x)\,dx\right]^r \left[\int_{x_i}^{x_{(n)}} f(x)\,dx\right]^s$$

$$f[x_{(1)}]\ f[x_{(n)}]\ f(x_i)\ dx_{(1)}\ dx_{(n)}\ dx_i. \tag{2-34}$$

The probability differential of the joint distribution of $[x_{(1)}, x_{(n)}, x_i]$ in the region under consideration is then given by

$$P[x_{(1)}, x_{(n)}, x_i]\ dx_{(1)}\ dx_{(n)}\ dx_i = \sum_{r=0}^{n-3} p_r$$

$$= (n-1)!\ f[x_{(1)}]\ f[x_{(n)}]\ f(x_i)\ dx_{(1)}\ dx_{(n)}\ dx_i$$

$$\sum_{r+s=n-3} \frac{1}{r!s!}\left[\int_{x_{(1)}}^{x_i} f(x)\,dx\right]^r \left[\int_{x_i}^{x_{(n)}} f(x)\,dx\right]^s$$

$$= \frac{(n-1)!}{(n-3)!}\ f[x_{(1)}]\ f[x_{(n)}] f(x_i)\ dx_{(1)}\ dx_{(n)}\ dx_i$$

$$\sum_{r+s=n-3} \frac{(n-3)!}{r!s!}\left[\int_{x_{(1)}}^{x_i} f(x)\,dx\right]^r \left[\int_{x_i}^{x_{(n)}} f(x)\,dx\right]^s$$

$$= (n-1)(n-2)\ f[x_{(1)}]\ f[x_{(n)}]\ f(x_i)\ dx_{(1)}\ dx_{(n)}\ dx_i$$

$$\left[\int_{x_{(1)}}^{x_i} f(x)\,dx + \int_{x_i}^{x_{(n)}} f(x)\,dx\right]^{n-3}$$

$$= (n - 1)(n - 2)\ f[x_{(1)}]\ f[x_{(n)}]\ f(x_i) \left[\int_{x_{(1)}}^{x_{(n)}} f(x)\ dx\right]^{n-3}$$

$$dx_{(1)}\ dx_{(n)}\ dx_i.$$

Hence

$$P[x_{(1)}, x_{(n)}, x_i] = (n - 1)(n - 2)\ f[x_{(1)}]\ f[x_{(n)}]\ f(x_i) \left[\int_{x_{(1)}}^{x_{(n)}} f(x)\ dx\right]^{n-3}$$

$$[x_{(1)} < x_i < x_{(n)},\ n > 2] \qquad (2\text{-}35)$$

It may also be actually verified by integration of Eq. (2-35) with respect to $x_{(1)}$, $x_{(n)}$, and x_i that the probability measure of Region 1 is $(n - 2)/n$.

REGION 2: $x_{(1)} = x_i < x_{(n)}$.

This is a two-dimensional region of probability measure $1/n$ in the three-dimensional space of $[x_{(1)}, x_{(n)}, x_i]$, and hence the three-dimensional joint distribution of $[x_{(1)}, x_{(n)}, x_i]$ is degenerate in this region. This distribution can be regarded as a nondegenerate two-dimensional distribution in the space of $[x_{(1)}, x_{(n)}]$, whose p.d.f. will be denoted by the symbol $P[x_{(1)}, x_{(n)}, x_i = x_{(1)}]$, and the probability differential will be denoted by $P[x_{(1)}, x_{(n)}, x_i = x_{(1)}]\ dx_{(1)}\ dx_{(n)}$. We note that $P[x_{(1)}, x_{(n)}, x_i = x_{(1)}]\ dx_{(1)}\ dx_{(n)}$ is precisely the probability that x_i falls in the range $[x_{(1)} - \frac{1}{2}\ dx_{(1)}, x_{(1)} + \frac{1}{2}\ dx_{(1)}]$; and of the remaining $n - 1$ sample members, $n - 2$ fall in the range $[x_{(1)} + \frac{1}{2}\ dx_{(1)}, x_{(n)} - \frac{1}{2}\ dx_{(n)}]$ and just one falls in the range $[x_{(n)} - \frac{1}{2}\ dx_{(n)}, x_{(n)} + \frac{1}{2}\ dx_{(n)}]$. Hence

$$P[x_{(1)}, x_{(n)}, x_i = x_{(1)}] = \frac{(n - 1)!}{(n - 2)!}\ f[x_{(1)}]\ f[x_{(n)}] \left[\int_{x_{(1)}}^{x_{(n)}} f(x)\ dx\right]^{n-2},$$

that is,

$$P[x_{(1)}, x_{(n)}, x_i = x_{(1)}] = (n-1)\ f[x_{(1)}]\ f[x_{(n)}] \left[\int_{x_{(1)}}^{x_{(n)}} f(x)\ dx\right]^{n-2}$$

$$[x_{(1)} = x_i < x_{(n)},\ n > 2]. \qquad (2\text{-}36)$$

By actual integration of Eq. (2-36) with respect to $x_{(1)}$ and $x_{(n)}$, it will be seen that the probability measure of Region 2 is $1/n$.

REGION 3: $x_{(1)} < x_i = x_{(n)}$.

The argument here is exactly similar to that used in Region 2, and

$$P[x_{(1)}, x_{(n)}, x_i = x_{(n)}] = (n-1)\ f[x_{(1)}]\ f[x_{(n)}] \left[\int_{x_{(1)}}^{x_{(n)}} f(x)\ dx\right]^{n-2}$$

$$[x_{(1)} < x_i = x_{(n)},\ n > 2]. \qquad (2\text{-}37)$$

REGION 4: $x_{(1)} = x_i = x_{(n)}$.

This is a one-dimensional region of probability measure zero in the three-dimensional space of $[x_{(1)}, x_{(n)}, x_i]$. If we treat the distribution in this region as a one-dimensional distribution in the space of one of the variables, say x_i, the p.d.f. of the distribution will be zero for all possible values of x_i. Denoting the p.d.f. of this distribution by $P[x_i, x_i = x_{(1)} = x_{(n)}]$, we have

$$P[x_i, x_{(1)} = x_i = x_{(n)}] = 0 \qquad [x_{(1)} = x_i = x_{(n)},\ n > 2]. \qquad (2\text{-}38)$$

Since the probability measure of Region 4 is zero, this region can be amalgamated with either Region 2 or Region 3. However, we prefer to treat this region separately to preserve the mutually exclusive character of the four regions, as well as to maintain symmetry of notation.

Hence Eqs. (2-35) - (2-38) give the p.d.f. of the joint distribution of $[x_{(1)}, x_{(n)}, x_i]$ for the case $n > 2$. We now consider Case II.

CASE II: $n = 2$.

Here, since x_i ($i = 1$ or 2) must be either $x_{(1)}$ or $x_{(2)}$, or all three may be equal, we divide the region of possible values of $[x_{(1)}, x_{(2)}, x_i]$ into three mutually exclusive regions.

REGION 1: $x_{(1)} = x_i < x_{(2)}$.

Here $P[x_{(1)}, x_{(2)}, x_i = x_{(1)}]\, dx_{(1)}\, dx_{(n)}$ is the probability that in a random sample of two members, x_i falls in the range $[x_{(1)} - \frac{1}{2} dx_{(1)}, x_{(1)} + \frac{1}{2} dx_{(1)}]$ and the remaining member falls in the range $[x_{(2)} - \frac{1}{2} dx_{(2)}, x_{(2)} + \frac{1}{2} dx_{(2)}]$. Hence

$$P[x_{(1)}, x_{(2)}, x_i = x_{(1)}] = f[x_{(1)}]\, f[x_{(2)}] \quad [x_{(1)} = x_i < x_{(2)}, n = 2]. \tag{2-39}$$

The probability measure of this region is ½.

REGION 2: $x_{(1)} < x_i = x_{(2)}$.

As in Region 1,

$$P[x_{(1)}, x_{(2)}, x_i = x_{(2)}] = f[x_{(1)}]\, f[x_{(2)}] \quad [x_{(1)} < x_i = x_{(2)}, n = 2]. \tag{2-40}$$

The probability measure of this region is ½.

REGION 3: $x_{(1)} = x_i = x_{(2)}$.

In this region

$$P[x_i, x_{(1)} = x_i = x_{(2)}] = 0 \quad [x_{(1)} = x_i = x_{(2)}, n = 2]. \tag{2-41}$$

The probability measure of this region is zero. Equations (2-39) - (2-41) give the p.d.f. of the joint sampling distribution of $[x_{(1)}, x_{(n)}, x_i]$ for the case $n = 2$.

It is verified that if we write $n = 2$ in Eqs. (2-36) - (2-38) of the case $n > 2$, we get Eqs. (2-39) - (2-41) respectively of the case $n = 2$. Hence the p.d.f. of the joint sampling distribution of $[x_{(1)}, x_{(n)}, x_i]$ when $n \geq 2$ is given by Eqs. (2-42) - (2-45), which follow.

$$P[x_{(1)}, x_{(n)}, x_i] = (n-1)(n-2)\ f[x_{(1)}]f[x_{(n)}]f(x_i) \left[\int_{x_{(1)}}^{x_{(n)}} f(x)\ dx\right]^{n-3}$$

$$[x_{(1)} < x_i < x_{(n)},\ n > 2]. \qquad (2\text{-}42)$$

$$P[x_{(1)}, x_{(n)}, x_i = x_{(1)}] = (n-1)\ f[x_{(1)}]\ f[x_{(n)}] \left[\int_{x_{(1)}}^{x_{(n)}} f(x)\ dx\right]^{n-2}$$

$$[x_{(1)} = x_i < x_{(n)},\ n \geq 2]. \qquad (2\text{-}43)$$

$$P[x_{(1)}, x_{(n)}, x_i = x_{(n)}] = (n-1)\ f[x_{(1)}]\ f[x_{(n)}] \left[\int_{x_{(1)}}^{x_{(n)}} f(x)\ dx\right]^{n-2}$$

$$[x_{(1)} < x_i = x_{(n)},\ n \geq 2]. \qquad (2\text{-}44)$$

$$P[x_i, x_{(1)} = x_i = x_{(n)}] = 0 \qquad [x_{(1)} = x_i = x_{(n)},\ n \geq 2]. \qquad (2\text{-}45)$$

2.6.1 The joint sampling distribution of $[x_{(1)}, x_{(n)}]$ when $n \geq 2$ for a distribution with p.d.f. f(x)

The joint sampling distribution of $[x_{(1)}, x_{(n)}]$ when $n \geq 2$ is well-known, and the p.d.f. of this distribution is given by

$$P[x_{(1)}, x_{(n)}] = n(n-1)\ f[x_{(1)}]\ f[x_{(n)}] \left[\int_{x_{(1)}}^{x_{(n)}} f(x)\ dx\right]^{n-2}. \qquad (2\text{-}46)$$

As a useful check on the results of Sec. 2.6, it may be noted that Eq. (2-46) is also deducible from Eqs. (2-42) - (2-45) by integration with respect to x_i.

2.6.2 The conditional distribution of x_i for a fixed $[x_{(1)}, x_{(n)}]$ when $n \geq 2$ for a distribution with p.d.f. f(x)

From Eqs. (2-42) - (2-46) we can obtain the conditional distribution of x_i for a fixed $[x_{(1)}, x_{(n)}]$ when $n \geq 2$. First we note that the range of the conditional distribution of x_i for a fixed $[x_{(1)}, x_{(n)}]$ is the closed interval $[x_{(1)}, x_{(n)}]$. We make the following cases:

CASE I: $x_{(1)} < x_i < x_{(n)}$, $n > 2$.

Here the conditional distribution of x_i for a fixed $[x_{(1)}, x_{(n)}]$ is nondegenerate with p.d.f. $P[x_i|x_{(1)}, x_{(n)}]$ given by

$$P[x_i|x_{(1)}, x_{(n)}] = \frac{P[x_{(1)}, x_{(n)}, x_i]}{P[x_{(1)}, x_{(n)}]}. \tag{2-47}$$

Using Eqs. (2-42) and (2-46) in Eq. (2-47), we have

$$P[x_i|x_{(1)}, x_{(n)}] = \frac{n-2}{n} \frac{f(x_i)}{\int_{x_{(1)}}^{x_{(n)}} f(x)\, dx}$$

$$[x_{(1)} < x_i < x_{(n)}, n > 2]. \tag{2-48}$$

CASE II: $x_{(1)} = x_i < x_{(n)}$, $n \geq 2$.

Here the point $x_i = x_{(1)}$ is a "discrete mass point" (See Cramer, 1946, for a definition of this term) of the conditional distribution of x_i for a fixed $[x_{(1)}, x_{(n)}]$, and we use the symbol $P[x_i = x_{(1)}|x_{(1)}, x_{(n)}]$ to denote the conditional probability at the point $x_i = x_{(1)}$. We have

$$P[x_i = x_{(1)}|x_{(1)}, x_{(n)}] = \frac{P[x_{(1)} = x_i < x_{(n)}]}{P[x_{(1)}, x_{(n)}]}. \tag{2-49}$$

Using Eqs. (2-43) and (2-46) in Eq. (2-49), we have

$$P[x_i = x_{(1)}|x_{(1)}, x_{(n)}] = \frac{1}{n} \qquad [x_{(1)} = x_i < x_{(n)}, n \geq 2]. \tag{2-50}$$

CASE III: $x_{(1)} < x_i = x_{(n)}$, $n \geq 2$.

The point $x_i = x_{(n)}$ is a discrete mass point of the conditional distribution of x_i for a fixed $[x_{(1)}, x_{(n)}]$. Equations (2-44) and (2-46) give

$$P[x_i = x_{(n)}|x_{(1)}, x_{(n)}] = \frac{1}{n} \qquad [x_{(1)} < x_i = x_{(n)}, n \geq 2]. \tag{2-51}$$

It should be noted that since

$$P[x_{(1)}, x_{(n)}] = 0 \quad \text{when} \quad x_{(1)} = x_{(n)},$$

the conditional distribution of x_i for a fixed $[x_{(1)}, x_{(n)}]$ is not defined at the point $x_i = x_{(1)} = x_{(n)}$, even though Eq. (2-45) gives

$$P[x_i, x_{(1)} = x_i = x_{(n)}] = 0 \qquad [x_{(1)} = x_i = x_{(n)}, \; n \geq 2].$$

Eqs. (2-48), (2-50) and (2-51) give the conditional distribution of x_i for a fixed $[x_{(1)}, x_{(n)}]$ for $n \geq 2$, for a distribution with p.d.f. $f(x)$. It may be noted that the endpoints of the range $[x_{(1)}, x_{(n)}]$ of the conditional distribution of x_i for a fixed $[x_{(1)}, x_{(n)}]$ are discrete mass points each with probability $1/n$, and the remaining probability $(n - 2)/n$ is distributed over the open interval $[x_{(1)}, x_{(n)}]$.

2.7 Sets Which are Mutually Exclusive "In Probability"

If sets $S_1, \ldots, S_n$ are such that the probability measure of the product of any two of the sets is zero, we shall call the sets mutually exclusive "in probability."

We have $P(S_i S_j) = 0 \qquad (i, j = 1, \ldots, n, \quad i \neq j)$.

Such sets may not be mutually exclusive according to the usual definition, but they obey the addition rule of probabilities for mutually exclusive sets:

$$P(S_1 + \ldots + S_n) = P(S_1) + \ldots + P(S_n). \tag{2-52}$$

If the definitions of the sets $S_1, \ldots, S_n$ depend on the differential dx of a variable x, and if to order dx we have

$$P(S_i S_j) = 0 \qquad (i, j = 1, \ldots, n, \; i \neq j),$$

we shall call the sets mutually exclusive "in probability" to order dx. The relation (2-52) holds good to order dx for such sets.

2.8 Definition of the Range of a Probability Distribution

Let $F(x)$ be the d.f. (distribution function) of a one-dimensional probability distribution. Although $F(x)$ is defined over the interval $R_1 = (-\infty, \infty)$, and although R_1 is often spoken of as the range of the probability distribution, we shall use the term "range of a probability distribution" in the sense of the smallest closed interval (a, b) having the property

$$P(a, b) = 1, \tag{2-53}$$

where the symbol $P(a, b)$ denotes the probability measure of the closed interval (a, b). In special cases either a may be $-\infty$ or b may be $+\infty$ or both may hold good. It can be easily shown that the range (a, b) as defined by Eq. (2-53) is unique. Further, a is given by

$$\begin{aligned} & F(x) = 0 \quad \text{for all } x < a, \\ \text{and} \quad & F(x) > 0 \quad \text{for all } x > a, \end{aligned} \tag{2-54}$$

and b is given by

$$\begin{aligned} & F(x) < 1 \quad \text{for all } x < b, \\ \text{and} \quad & F(x) = 1 \quad \text{for all } x > b. \end{aligned} \tag{2-55}$$

If the distribution possesses p.d.f. $f(x)$, then a is given by

$$\begin{aligned} & f(x) = 0 \quad \text{for all } x < a, \\ \text{and} \quad & f(x) > 0 \quad \text{for at least one } x \text{ belonging to the neighborhood } (a, a + \varepsilon) \text{ of } a, \text{ for every } \varepsilon > 0. \end{aligned} \tag{2-56}$$

Similarly b is given by

$$\begin{aligned} & f(x) = 0 \quad \text{for all } x > b, \\ \text{and} \quad & f(x) > 0 \quad \text{for at least one } x \text{ belonging to the neighborhood } (b - \varepsilon, b) \text{ of } b, \text{ for every } \varepsilon > 0. \end{aligned} \tag{2-57}$$

2.9 Definitions of the Regular and Nonregular Cases of Estimation of Unknown Parameters

Consider a family of distributions depending on some unknown parameters. If the ranges (as defined in Sec. 2.8) of the family of distributions are independent of the parameters (i.e., the range is the same for every member of the family of distributions), we shall call it a regular case of estimation. If the ranges depend on the parameters, we shall call it a nonregular case of estimation.

2.10 Definitions of Pure-Type and Mixed-Type Cases in Search for Sufficient Statistics for Parameters in Nonregular Cases

A nonregular case admitting jointly sufficient statistics for all parameters will be said to be of the "pure type" if every member of the minimal set of sufficient statistics is a function of only the smallest and the greatest members in the sample.

A nonregular case admitting jointly sufficient statistics for all parameters will be said to be of the "mixed type" if it is not of the pure type.

3. The Nonregular Case with a Single Sufficient Statistic for a Single Unknown Parameter

3.1 Statement of the Problem and the Assumptions

We investigate the general form of the p.d.f. $f(x; \theta)$ of the family of absolutely continuous univariate distributions depending on a single unknown parameter θ which admits a single sufficient statistic for all samples of any size $n \geq 2$, when the ranges $[a(\theta), b(\theta)]$ of the family of distributions depend on θ. We assume that the range of admissible values of the parameter θ is some nondegenerate interval $\Omega = (\lambda_1, \lambda_2)$ which may be closed, open, half-open, finite, or infinite. $f(x; \theta) = 0$ for all $x < a(\theta)$, for all $x > b(\theta)$, and for all $\theta \in \Omega$. Since the case is nonregular, at least one terminal of the range (a, b) must depend on θ. Hence at least one of the two functions $a(\theta)$ and $b(\theta)$ is not constant for all $\theta \in \Omega$. We make the following assumptions about $a(\theta)$ and $b(\theta)$:

1. For all $\theta \in \Omega$, $a(\theta)$ and $b(\theta)$ are monotonic functions of θ in opposite senses; i.e., either $a(\theta)$ is m.i. and $b(\theta)$ is m.d., or else $a(\theta)$ is m.d. and $b(\theta)$ is m.i.

2. For all $\theta \in \Omega$, $a(\theta)$ and $b(\theta)$ are continuous and differentiable functions of θ.

3. For all $\theta \in \Omega$, $a(\theta) < b(\theta)$.

In assumption 3, we have excluded the value of θ, if it exists, which makes $a(\theta) = b(\theta)$ to avoid the trivial degenerate distribution, which is also contrary to our assumption that the distributions considered are absolutely continuous. If $a(\theta)$ is m.d. and $b(\theta)$ is m.i., we shall briefly say that "$a(\theta)$ and $b(\theta)$ are monotonic Type I," and if $a(\theta)$ is m.i. and $b(\theta)$ is m.d., we shall say that "$a(\theta)$ and $b(\theta)$ are monotonic Type II."

Incidentally it may be remarked here in connection with assumption 2 that, since $\int_{a(\theta)}^{b(\theta)} f(x;\ \theta)\, dx = 1$ for all $\theta \in \Omega$, the assumption that one of the terminals $a(\theta)$ and $b(\theta)$ is a continuous and differentiable function of θ implies that the other terminal is also a continuous and differentiable function of θ almost everywhere in Ω.

3.2 Intervals of Values of $a(\theta)$ and $b(\theta)$ and the Union and Intersection Sets of Ranges of the Family of Distributions

Let I_a and I_b denote respectively the intervals of values assumed by $a(\theta)$ and $b(\theta)$ for all $\theta \in \Omega$. We have

$$I_a = [a(\lambda_2),\ a(\lambda_1)], \qquad I_b = [b(\lambda_1),\ b(\lambda_2)] \tag{3-1}$$

if $a(\theta)$ and $b(\theta)$ are monotonic Type I, and

$$I_a = [a(\lambda_1),\ a(\lambda_2)], \qquad I_b = [b(\lambda_2),\ b(\lambda_1)] \tag{3-2}$$

if $a(\theta)$ and $b(\theta)$ are monotonic Type II. In case Ω is open at an endpoint, say λ_1, the corresponding symbols $a(\lambda_1)$ and $b(\lambda_1)$ are supposed to denote limiting values as explained in Sec. 2.3. The intervals I_a and I_b may be closed, open, half-open, finite, or infinite. Further, one (but not both) of the intervals I_a and I_b

may be degenerate corresponding to the case of one (but not both) of the functions $a(\theta)$ and $b(\theta)$ being constant throughout Ω.

Since $a(\theta) < b(\theta)$ for all $\theta \varepsilon \Omega$, and $a(\theta)$ and $b(\theta)$ are monotonic in opposite senses, it follows that I_b is to the right side of I_a; I_a and I_b may have a common endpoint at which both are open. In Eqs. (3-1) and (3-2) we have

$$a(\lambda_1) \leq b(\lambda_1), \qquad a(\lambda_2) \leq b(\lambda_2), \qquad \max_{\theta\varepsilon\Omega} a(\theta) \leq \min_{\theta\varepsilon\Omega} b(\theta). \tag{3-3}$$

It should be noted that if an equality holds good in Eq. (3-3), it is an equality between limiting values, and Ω will be open at the corresponding endpoint λ_1 or λ_2.

Let the set ρ be the intersection of the ranges $[a(\theta), b(\theta)]$ for all $\theta \varepsilon \Omega$.

$$\rho = [\max_{\theta\varepsilon\Omega} a(\theta), \min_{\theta\varepsilon\Omega} b(\theta)]. \tag{3-4}$$

We have

$$\rho = [a(\lambda_1), b(\lambda_1)] \tag{3-5}$$

if $a(\theta)$ and $b(\theta)$ are monotonic Type I, and

$$\rho = [a(\lambda_2), b(\lambda_2)] \tag{3-6}$$

if $a(\theta)$ and $b(\theta)$ are monotonic Type II. We shall take ρ to be a closed interval by assigning to it the endpoints. It will be seen from Eqs. (3-1), (3-2), (3-5), and (3-6) that the interval ρ is bordered by the intervals I_a and I_b.

In our investigation we shall be confronted with two cases:

CASE I: $$\max_{\theta\varepsilon\Omega} a(\theta) = \min_{\theta\varepsilon\Omega} b(\theta). \tag{3-7}$$

This case arises when $a(\lambda_1) = b(\lambda_1)$ is the monotonic Type I case, and $a(\lambda_2) = b(\lambda_2)$ in the monotonic Type II case. In the case in point, the interval ρ is degenerate and consists of the single point $a(\lambda_1) = b(\lambda_1)$ in the monotonic Type I case, and the single point $a(\lambda_2) = b(\lambda_2)$ in the monotonic Type II case. When Eq. (3-7) holds good, we have for all $\theta \varepsilon \Omega$

$$a(\lambda_2) \leq a(\theta) < a(\lambda_1) = b(\lambda_1) < b(\theta) \leq b(\lambda_2) \tag{3-8}$$

for the monotonic Type I case, and

$$a(\lambda_1) \leq a(\theta) < a(\lambda_2) = b(\lambda_2) < b(\theta) \leq b(\lambda_1) \tag{3-9}$$

for the monotonic Type II case.

CASE II: $\max_{\theta\varepsilon\Omega} a(\theta) < \min_{\theta\varepsilon\Omega} b(\theta)$ (3-10)

This case arises when $a(\lambda_1) < b(\lambda_1)$ in the monotonic Type I case, and $a(\lambda_2) < b(\lambda_2)$ in the monotonic Type II case. The interval ρ is nondegenerate in Case II. When Eq. (3-10) holds good, we have for all $\theta \varepsilon \Omega$

$$a(\lambda_2) \leq a(\theta) \leq a(\lambda_1) < b(\lambda_1) \leq b(\theta) \leq b(\lambda_2) \tag{3-11}$$

for the monotonic Type I case, and

$$a(\lambda_1) \leq a(\theta) \leq a(\lambda_2) < b(\lambda_2) \leq b(\theta) \leq b(\lambda_1) \tag{3-12}$$

for the monotonic Type II case.

Let the set R be the union of the ranges $[a(\theta), b(\theta)]$ for all $\theta \varepsilon \Omega$,

$$R = [\min_{\theta\varepsilon\Omega} a(\theta), \max_{\theta\varepsilon\Omega} b(\theta)]. \tag{3-13}$$

We have $R = [a(\lambda_2), b(\lambda_2)]$ (3-14)

for the monotonic Type I case, and

$$R = [a(\lambda_1), b(\lambda_1)] \tag{3-15}$$

for the monotonic Type II case.

R may be closed or open at an endpoint according as I_a or I_b is closed or open at that endpoint.

Every random observation fron any member of the family of distributions under investigation belongs to R, and hence

$$P(x \varepsilon R) = 1 \quad \text{for all } \theta \varepsilon \Omega. \tag{3-16}$$

In special cases R may be equal to $R_1 = (-\infty, \infty)$. It may be noted that

$$R = I_a + \rho + I_b, \tag{3-17}$$

where the middle interval ρ is bordered on its left by I_a and on its right by I_b; ρ is closed; I_a and I_b have no common point; and every point of R belongs to one and only one of (I_a, ρ, I_b) except that the left endpoint of ρ may also belong to I_a, and the right endpoint of ρ may also belong to I_b.

3.3 Search for a Single Sufficient Statistic for θ when $n \geq 2$

In this section and in Secs. 3.4, 3.5, and 3.6 we shall consider the case when $a(\theta)$ and $b(\theta)$ are monotonic Type I. The results for the case when $a(\theta)$ and $b(\theta)$ are monotonic Type II follow on similar lines and will be described in Sec. 3.7. In the present case, we have

$$R = [a(\lambda_2), b(\lambda_2)] \tag{3-18}$$

$$= I_a + \rho + I_b, \tag{3-19}$$

where $I_a = [a(\lambda_2), a(\lambda_1)]$ (3-20)

$$\rho = [a(\lambda_1), b(\lambda_1)], \tag{3-21}$$

and $I_b = [b(\lambda_1), b(\lambda_2)].$ (3-22)

Let $x_1, \ldots, x_n$ be a random sample of n (≥ 2) independent observations from a distribution of the family with p.d.f. $f(x; \theta)$. If $x_{(1)}$ and $x_{(n)}$ are respectively the smallest and the greatest members in the sample, we have

$$a(\theta) \leq x_{(1)} \leq x_1, \ldots, x_n \leq x_{(n)} \leq b(\theta). \tag{3-23}$$

Now $a(\theta) \leq x_{(1)}$ is equivalent to $\theta \geq a_1^{-1}[x_{(1)}]$, (3-24)

and $b(\theta) \geq x_{(n)}$ is equivalent to $\theta \geq b_1^{-1}[x_{(n)}]$, (3-25)

where the functions a_1^{-1} and b_1^{-1} are as defined in Sec. 2.4. It should be noted that Eqs. (3-24) and (3-25) hold good for all $x_{(1)}$ and $x_{(n)}$ in the interval R, as shown in Sec. 2.3. Consider the single statistic T defined by

$$T = \max\,[a_1^{-1}\,[x_{(1)}],\, b_1^{-1}[x_{(n)}]]. \tag{3-26}$$

T is a single-valued function of $[x_{(1)}, x_{(n)}]$, and we note the following four situations:

1. If $x_{(1)} \in I_a$ and $x_{(n)} \in I_b$, we have $\lambda_1 \le a_1^{-1}[x_{(1)}] \le \lambda_2$ and $\lambda_1 \le b_1^{-1}[x_{(n)}] \le \lambda_2$. T is as given by Eq. (3-26).

2. If $x_{(1)} \in I_a$ and $x_{(n)} \notin I_b$, we have $b_1^{-1}[x_{(n)}] = \lambda_1$, and then Eq. (3-26) gives $T = a_1^{-1}[x_{(1)}]$.

3. If $x_{(1)} \notin I_a$ and $x_{(n)} \in I_b$, we have $a_1^{-1}[x_{(1)}] = \lambda_1$, and Eq. (3-26) gives $T = b_1^{-1}[x_{(n)}]$.

4. If $x_{(1)} \notin I_a$ and $x_{(n)} \notin I_b$, then $x_{(1)} \in \rho$, since $x_{(1)} \le x_{(n)}$. In this case $a_1^{-1}[x_{(1)}] = \lambda_1 = b_1^{-1}[x_{(n)}]$. Then Eq. (3-26) gives $T = \lambda_1$.

Incidentally, it may be pointed out here that in virtue of Eq. (2-18), a_1^{-1} has the same value λ_1 at the border point $a(\lambda_1)$ of I_a whether the border point belongs to I_a or to ρ or to both. Similarly b_1^{-1} has the same value at the border point $b(\lambda_1)$ of I_b, whether the border point belongs to I_b or to ρ or to both.

In virtue of Eqs. (3-24) - (3-26), we have

$$\theta \ge T \ge a_1^{-1}[x_{(1)}], \tag{3-27}$$

and $$\theta \ge T \ge b_1^{-1}[x_{(n)}]. \tag{3-28}$$

Since all values of each of $a_1^{-1}[x_{(1)}]$, $b_1^{-1}[x_{(n)}]$, and T lie in the extended closed interval Ω and since $a(\theta)$ and $b(\theta)$ continue to be m.d. and m.i. respectively in the extended closed interval Ω (see remarks in the first paragraph of Sec. 2.4), it follows from Eqs. (3-27) and (3-28) that

$$a(\theta) \le a(T) \le a\{a_1^{-1}[x_{(1)}]\} \tag{3-29}$$

and $b(\theta) \geq b(T) \geq b\{b_1^{-1}[x_{(n)}]\}$. (3-30)

Now consider the four situations mentioned above. We use formulae (2-6) and (2-16).

SITUATION 1

Here $a\{a_1^{-1}[x_{(1)}]\} = x_{(1)}$ and $b\{b_1^{-1}[x_{(n)}]\} = x_{(n)}$. Hence Eqs. (3-29) and (3-30) become respectively

$$a(\theta) \leq a(T) \leq x_{(1)} \tag{3-31}$$

and $b(\theta) \geq b(T) \geq x_{(n)}$. (3-32)

Equations (3-31) and (3-32) give

$$a(\theta) \leq a(T) \leq x_{(1)} \leq x_{(n)} \leq b(T) \leq b(\theta). \tag{3-33}$$

SITUATION 2

Here $a(T) = a\{a_1^{-1}[x_{(1)}]\} = x_{(1)}$ and $b\{b_1^{-1}[x_{(n)}]\} = b(\lambda_1) \geq x_{(n)}$, since $x_{(n)} \not\in I_b$. Hence Eqs. (3-29) and (3-30) give

$$a(\theta) \leq a(T) = x_{(1)} \leq x_{(n)} \leq b(\lambda_1) \leq b(T) \leq b(\theta). \tag{3-34}$$

SITUATION 3

Here $a\{a_1^{-1}[x_{(1)}]\} = a(\lambda_1) \leq x_{(1)}$ and $b(T) = b\{b_1^{-1}[x_{(n)}]\} = x_{(n)}$. Equations (3-29) and (3-30) give

$$a(\theta) \leq a(T) \leq a(\lambda_1) \leq x_{(1)} \leq x_{(n)} = b(T) \leq b(\theta). \tag{3-35}$$

SITUATION 4

Here $a(T) = a\{a_1^{-1}[x_{(1)}]\} = a(\lambda_1) \leq x_{(1)}$ and $b(T) = b\{b_1^{-1}[x_{(n)}]\} = b(\lambda_1) \geq x_{(n)}$. Equations (3-29) and (3-30) give

$$a(\theta) \leq a(T) = a(\lambda_1) \leq x_{(1)} \leq x_{(n)} \leq b(\lambda_1) = b(T) \leq b(\theta). \tag{3-36}$$

Equations (3-33) - (3-36) show that when T is fixed, each x_i lies in the range $[a(T), b(T)]$, which is independent of θ, for all sample points $(x_1, \ldots, x_n)$ $(n \geq 2)$ and for all $\theta \in \Omega$. Moreover, under the assumptions made, it will be seen from the manner in which T is defined that T (or any one-to-one function of T) is the only single statistic having this property, for all sample points $(x_1, \ldots, x_n)$ $(n \geq 2)$ and $\theta \in \Omega$. Hence, under the assumptions made, if a single sufficient statistic exists for θ for all random samples of any size $n \geq 2$ and for all $\theta \in \Omega$, it must be T as defined by Eq. (3-26), or any one-to-one function of T; and without any loss of generality, we may take a one-to-one function of T to be T itself. We have accordingly proved the following:

THEOREM 1

Under the assumptions of Sec. 3.1, where $a(\theta)$ and $b(\theta)$ are monotonic Type I, a necessary and sufficient condition for the existence of a single sufficient statistic for θ for all random samples of any size $n \geq 2$ and for all $\theta \in \Omega$ is that the single statistic T as defined by Eq. (3-26) should be sufficient for θ.

We now proceed to investigate the general form of the p.d.f. $f(x; \theta)$ of the family of absolutely continuous distributions, such that T [as defined by Eq. (3-26)] is a single sufficient statistic for θ for all random samples of any size $n \geq 2$ for all $\theta \in \Omega$. For this purpose we shall use the equivalent Form A of the definition of a single sufficient statistic (see Sec. 2.1). Form A requires derivation of the conditional distribution of x_i $(i = 1, \ldots, n)$ for a fixed T, and we shall obtain this conditional distribution in stages.

3.4 The Sampling Distribution of T as Defined by Equation (3-26), $\theta \geq T \geq b_1^{-1}[x_{(n)}]$, when $n > 2$

Under the assumptions of Sec. 3.1, and using Eq. (2-46), the p.d.f. of the joint sampling distribution of $[x_{(1)}, x_{(n)}]$ when $n \geq 2$ is

$$P[x_{(1)}, x_{(n)}; \theta] = n(n-1)\, f[x_{(1)}; \theta]\, f[x_{(n)}; \theta] \left[\int_{x_{(1)}}^{x_{(n)}} f(x; \theta)\, dx\right]^{n-2}$$

$$[a(\theta) \leq x_{(1)} \leq x_{(n)} \leq b(\theta)]. \qquad (3\text{-}37)$$

Let R_2 denote the two-dimensional region

$$a(\theta) \leq x_{(1)} \leq x_{(n)} \leq b(\theta) \tag{3-38}$$

in the space of $[x_{(1)}, x_{(2)}]$. We proceed to obtain the sampling distribution of T [as defined by Eq. (3-26)] from Eq. (3-37), when $n \geq 2$. First we note from Eqs. (3-27) and (3-28) that the upper terminal of the range of the sampling distribution of T is θ. The lower terminal of the range of T is λ_1. Hence the range of the distribution of T is the closed interval (λ_1, θ).

First we shall prove two lemmas concerning the sampling distribution of T as defined by Eq. (3-26).

LEMMA 1

$T = \lambda_1$ if, and only if, $x_{(1)} \varepsilon \rho$ and $x_{(n)} \varepsilon \rho$.

Proof. $T = \lambda_1$ if, and only if,

$$\max\{a_1^{-1}[x_{(1)}], b_1^{-1}[x_{(n)}]\} = \lambda_1. \tag{3-39}$$

Since $\lambda_1 \leq a_1^{-1}[x_{(1)}] \leq \lambda_2$ and $\lambda_1 \leq b_1^{-1}[x_{(n)}] \leq \lambda_2$, we have $T = \lambda_1$ if, and only if

$$a_1^{-1}[x_{(1)}] = \lambda_1 \tag{3-40}$$

and $$b_1^{-1}[x_{(n)}] = \lambda_1. \tag{3-41}$$

First consider Eq. (3-40). If $x_{(1)} \varepsilon I_a$, we have [in virtue of Eq. (2-6)] $x_{(1)} = a(\lambda_1)$. If $x_{(1)} \not\varepsilon I_a$, we have $x_{(1)} \geq a(\lambda_1)$. Hence Eq. (3-40) implies $x_{(1)} \geq a(\lambda_1)$.

Similarly Eq. (3-41) implies $x_{(n)} \leq b(\lambda_1)$. Since $x_{(1)} \leq x_{(n)}$, both Eqs. (3-40) and (3-41) imply $a(\lambda_1) \leq x_{(1)} \leq x_{(n)} \leq b(\lambda_1)$, i.e., $x_{(1)} \varepsilon \rho$ and $x_{(n)} \varepsilon \rho$.

We shall find it convenient to use the terms "discrete mass point" and "continuity point" (Cramer, 1946) in Lemma 2, which follows.

LEMMA 2

The sampling distribution of T has a discrete mass point if, and only if,

$$P_\theta(\rho) = \int_{a(\lambda_1)}^{b(\lambda_1)} f(x;\ \theta)\ dx > 0, \tag{3-42}$$

in which case $T = \lambda_1$ is the only discrete mass point of the sampling distribution of T containing the probability

$$P(T = \lambda_1) = [P_\theta(\rho)]^n.$$

Proof. First we note that $P_\theta(\rho)$ is the probability measure of the interval $\rho = [a(\lambda_1), b(\lambda_1)]$ with respect to the p.d.f. $f(x;\ \theta)$. The condition $P_\theta(\rho) = 0$ is satisfied either if ρ is degenerate or if ρ does not contain any mass distribution, which condition is satisfied if $f(x;\ \theta) = 0$ for almost all $x \in \rho$ and all $\theta \in \Omega$.

First consider any value of T, say $T_0 \neq \lambda_1$. Then $T_0 > \lambda_1$.

$$\begin{aligned} \text{Then } P(T = T_0) &= P(\max\{a_1^{-1}[x_{(1)}],\ b_1^{-1}[x_{(n)}]\} = T_0) \\ &= P\{\text{either } T_0 = a_1^{-1}[x_{(1)}] > b_1^{-1}[x_{(n)}] \\ &\qquad \text{or } T_0 = b_1^{-1}[x_{(n)}] > a_1^{-1}[x_{(1)}] \\ &\qquad \text{or } T_0 = a_1^{-1}[x_{(1)}] = b_1^{-1}[x_{(n)}]\} \\ &= P\{T_0 = a_1^{-1}[x_{(1)}] > b_1^{-1}[x_{(n)}]\} \\ &\quad + P\{T_0 = b_1^{-1}[x_{(n)}] > a_1^{-1}[x_{(1)}]\} \\ &\quad + P\{T_0 = a_1^{-1}[x_{(1)}] = b_1^{-1}[x_{(n)}]\}. \end{aligned} \tag{3-43}$$

Now since the statement $T_0 = a_1^{-1}[x_{(1)}] > b_1^{-1}[x_{(n)}]$ implies the statement $T_0 = a_1^{-1}[x_{(1)}]$,

$$P\{T_0 = a_1^{-1}[x_{(1)}] > b_1^{-1}[x_{(n)}]\} \le P\{T_0 = a_1^{-1}[x_{(1)}]\}. \tag{3-44}$$

Also, since $T_0 = a_1^{-1}[x_{(1)}] > \lambda_1$, $x_{(1)} \varepsilon I_{a}$, and therefore $a(T_0) = a\{a_1^{-1}[x_{(1)}]\} = x_{(1)}$. Hence $T_0 = a_1^{-1}\ x_{(1)}$ implies $x_{(1)} = a(T_0)$, and therefore

$$P\{T_0 = a_1^{-1}[x_{(1)}]\} \leq P[x_{(1)} = a(T_0)]. \tag{3-45}$$

$$\text{Hence } P\{T_0 = a_1^{-1}[x_{(1)}] > b_1^{-1}[x_{(n)}]\} \leq P[x_{(1)} = a(T_0)]. \tag{3-46}$$

$$\text{Similarly } P\{T_0 = b_1^{-1}[x_{(1)}] > a_1^{-1}[x_{(n)}]\} \leq P[x_{(n)} = b(T_0)]. \tag{3-47}$$

Also, since $T_0 = a_1^{-1}[x_{(1)}] = b_1^{-1}[x_{(n)}]$ implies $T_0 = a_1^{-1}[x_{(1)}]$, which implies $x_{(1)} = a(T_0)$ (since $T_0 > \lambda_1$),

$$P\{T_0 = a_1^{-1}[x_{(1)}] = b_1^{-1}[x_{(n)}]\} \leq P[x_{(1)} = a(T_0)]. \tag{3-48}$$

Hence Eqs. (3-43) and (3-46) - (3-48) give

$$P(T = T_0) \leq 2P[x_{(1)} = a(T_0)] + P[x_{(n)} = b(T_0)]. \tag{3-49}$$

Now since the given family of distributions possess p.d.f. $f(x;\ \theta)$, the sampling distribution of each of $x_{(1)}$ and $x_{(n)}$ has no discrete mass point.

$$\text{Hence } P[x_{(1)} = a(T_0)] = 0 = P[x_{(n)} = b(T_0)]. \tag{3-50}$$

Then Eqs. (3-49) and (3-50) give

$$P(T = T_0) = 0 \qquad (T_0 \neq \lambda_1). \tag{3-51}$$

Thus $T_0 (T_0 \neq \lambda_1)$ is a continuity point of the sampling distribution of T.

Now consider the value $T = \lambda_1$. By Lemma 1, $T = \lambda_1$ if, and only if, $x_{(1)} \varepsilon \rho$ and $x_{(n)} \varepsilon \rho$. Hence

$$P(T = \lambda_1) = P[x_{(1)} \varepsilon \rho \text{ and } x_{(n)} \varepsilon \rho] = [P_\theta(\rho)]^n, \tag{3-52}$$

where $P_\theta(\rho)$ is given by Eq. (3-42).

Since $P_\theta(\rho) \geq 0$, it follows from Eq. (3-52) that λ_1 is a discrete mass point of the sampling distribution of T if, and only if, $P_\theta(\rho) > 0$, in which case $P(T = \lambda_1) = [P_\theta(\rho)]^n$.

Remark on Lemma 2. In virtue of the results of Lemma 2, we need first obtain the p.d.f. $P(T; \theta)$ of the sampling distribution of T for the half-open interval $(\lambda_1 < T \leq \theta)$, i.e., excluding the value $T = \lambda_1$. The same p.d.f. $P(T; \theta)$ will represent the p.d.f. of T for the closed interval $(\lambda_1 \leq T \leq \theta)$ also, if $P_\theta(\rho) = 0$. If $P_\theta(\rho) \neq 0$, complete the distribution of T by the additional statement (3-52), which gives the probability at the discrete mass point λ_1 when $P_\theta(\rho) \neq 0$.

It should be noted that the p.d.f. $P(T; \theta)$ will represent both the cases $P_\theta(\rho) = 0$ and $P_\theta(\rho) \neq 0$ for the half-open interval $(\lambda_1 < T \leq \theta)$.

To obtain the sampling distribution of T, consider any value of T in the half-open interval $(\lambda_1 < T \leq \theta)$. Consider the set Z of points $(x_{(1)}, x_{(n)})$ $[x_{(1)} < x_{(n)}]$ in the two-dimensional region R_2 [Eq. (3-38)] in the space of $[x_{(1)}, x_{(n)}]$, which give the given value T of the statistic T. In view of the definition (3-26) of T, we have

$$Z = Z_1 + Z_2 + Z_3, \tag{3-53}$$

where the subsets Z_1, Z_2, and Z_3 are mutually exclusive and consist of points $[x_{(1)}, x_{(n)}]$ defined as follows:

$$Z_1: \quad T = a_1^{-1}[x_{(1)}] > b_1^{-1}[x_{(n)}]; \tag{3-54}$$

$$Z_2: \quad T = b_1^{-1}[x_{(n)}] > a_1^{-1}[x_{(1)}]; \tag{3-55}$$

$$Z_3: \quad T = b_1^{-1}[x_{(n)}] = a_1^{-1}[x_{(1)}]. \tag{3-56}$$

First consider Z_1. Since $T = a_1^{-1}[x_{(1)}] > \lambda_1$, $x_{(1)} \in I_a$. Hence $T = a_1^{-1}[x_{(1)}]$ implies $a(T) = x_{(1)}$. If $x_{(n)} \in I_b$, $T > b_1^{-1}[x_{(n)}]$ implies $b(T) \geq x_{(n)}$, and we have

$$a(T) = x_{(1)} \qquad \text{and} \qquad a(T) \leq x_{(n)} \leq b(T).$$

If $x_{(n)} \notin I_b$, we have $x_{(n)} \leq b(\lambda_1)$ and $b_1^{-1}[x_{(n)}] = \lambda_1$. Then $T > b_1^{-1}[x_{(n)}]$ implies $b(T) \geq b(\lambda_1) \geq x_{(n)}$, i.e., $T > b_1^{-1}[x_{(n)}]$

implies $b(T) \geq x_{(n)}$. Hence Eq. (3-54) implies

$$a(T) = x_{(1)}, \quad \text{and} \quad a(T) \leq x_{(n)} \leq b(T),$$

i.e., $Z_1 \subset W_1$ (the symbol $\subset$ means "is a subset of"), where the set W_1 is defined by

$$W_1: \quad a(T) = x_{(1)}, \quad a(T) \leq x_{(n)} \leq b(T). \tag{3-57}$$

Similarly $Z_2 \subset W_2$ and $Z_3 \subset W_3$, where

$$W_2: \quad a(T) \leq x_{(1)} \leq b(T), \quad x_{(n)} = b(T); \tag{3-58}$$

$$W_3: \quad a(T) = x_{(1)}, \quad x_{(n)} = b(T). \tag{3-59}$$

It may be noted that since $t \neq \lambda_1$, $a(T) < b(T)$.

$$\text{Write } W_1 + W_2 + W_3 = V_1 + V_2 + V_3 = V, \tag{3-60}$$

where the sets V_1, V_2, and V_3 are mutually exclusive and given by

$$V_1: \quad a(T) = x_{(1)}, \quad a(T) \leq x_{(n)} < b(T); \tag{3-61}$$

$$V_2: \quad a(T) < x_{(1)} \leq b(T), \quad x_{(n)} = b(T); \tag{3-62}$$

$$V_3: \quad a(T) = x_{(1)}, \quad x_{(n)} = b(T). \tag{3-63}$$

It follows that

$$Z = Z_1 + Z_2 + Z_3 \subset W_1 + W_2 + W_3 = V. \tag{3-64}$$

In each of Eqs. (3-61) - (3-63), where there is an equation in T, T is supposed (by definition) to be the smallest root satisfying the equation. Conversely, it is easily shown that every point of V is a point of either Z_1 or Z_2 or Z_3.

$$\text{Hence } V \subset Z_1 + Z_2 + Z_3 = Z. \tag{3-65}$$

Hence from Eqs. (3-64) and (3-65) we have

$$V = Z, \tag{3-66}$$

and hence $V = V_1 + V_2 + V_3$ is the set of points $[x_{(1)}, x_{(n)}]$ which give a given value T of the statistic T, where V_1, V_2, and V_3 are defined by Eqs. (3-61) - (3-63).

Now let S be the set of points $[x_{(1)}, x_{(n)}]$ in the region R_2 which give a value of T in the interval $(T - \frac{1}{2}\,dT,\ T + \frac{1}{2}\,dT)$. It follows from Eqs. (3-61) - (3-63) that

$$S = S_1 + S_2 + S_3, \tag{3-67}$$

where S_1, S_2, and S_3 are sets of points $[x_{(1)}, x_{(n)}]$ defined as follows:

$$S_1:\quad a(T + \tfrac{1}{2}\,dT) \le x_{(1)} \le a(T - \tfrac{1}{2}\,dT), \qquad \ell_1 \le x_{(n)} < \ell_2, \tag{3-68}$$

where ℓ_1 and ℓ_2 are any numbers satisfying

$$a(T + \tfrac{1}{2}\,dT) \le \ell_1 \le a(T - \tfrac{1}{2}\,dT),\ b(T - \tfrac{1}{2}\,dT) \le \ell_2 \le b(T + \tfrac{1}{2}\,dT).$$

$$S_2:\quad \ell_1 < x_{(1)} \le \ell_2, \qquad b(T - \tfrac{1}{2}\,dT) \le x_{(n)} \le b(T + \tfrac{1}{2}\,dT), \tag{3-69}$$

where ℓ_1 and ℓ_2 are as defined as in S_1.

$$S_3:\quad a(T + \tfrac{1}{2}\,dT) \le x_{(1)} \le a(T - \tfrac{1}{2}\,dT),$$

$$b(T - \tfrac{1}{2}\,dT) \le x_{(n)} \le b(T + \tfrac{1}{2}\,dT). \tag{3-70}$$

It may be remembered in the above expressions that $a(\theta)$ is m.d. and $b(\theta)$ is m.i. It may also be noted that in any integral over (ℓ_1, ℓ_2), we may approximately replace ℓ_1 by $a(T)$, and ℓ_2 by $b(T)$.

The probability measures of the sets S_1, S_2, and S_3 to order dT are:

$$P(S_1) = -n(n-1)\ f[a(T);\ \theta]\ a'(T)\ dT \int_{a(T)}^{b(T)} f(x_n;\ \theta) \left[\int_{a(T)}^{x_{(n)}} f(x;\ \theta)\ dx\right]^{n-2} dx_{(n)}$$

$$= -n(n-1)\ f[a(T);\ \theta]\ a'(T)\ dT \left[\frac{1}{n-1}\left(\int_{a(T)}^{x_{(n)}} f(x;\ \theta)\ dx\right)^{n-1}\right]_{x_{(n)} = a(T)}^{x_{(n)} = b(T)}$$

$$= -n\ f[a(T);\ \theta]\ a'(T) \left\{\int_{a(T)}^{b(T)} f(x;\ \theta)\ dx\right\}^{n-1} dT, \qquad (3\text{-}71)$$

where the prime denotes differentiation.

$$P(S_2) = n(n-1)\ f[b(T);\ \theta]\ b'(T)\ dT$$

$$\times \int_{a(T)}^{b(T)} f[x_{(1)};\ \theta] \left[\int_{x_{(1)}}^{b(T)} f(x;\ \theta)\ dx\right]^{n-2} dx_{(1)}$$

$$= n(n-1)\ f[b(T);\ \theta]\ b'(T)\ dT$$

$$\left[-\frac{1}{n-1}\left\{\int_{x_{(1)}}^{b(T)} f(x;\ \theta)\ dx\right\}^{n-1}\right]_{x_{(1)} = a(T)}^{x_{(1)} = b(T)}$$

$$= n\ f[b(T);\ \theta]\ b'(T) \left\{\int_{a(T)}^{b(T)} f(x;\ \theta)\ dx\right\}^{n-1} dT. \qquad (3\text{-}72)$$

$$P(S_3) = -n(n-1)\ f[a(T);\ \theta]\ a'(T)\ dT$$

$$f[b(T);\ \theta]\ b'(T)\ dT \cdot \left\{\int_{a(T)}^{b(T)} f(x;\ \theta)\ dx\right\}^{n-2}$$

$$= 0 \quad \text{to order } dT. \qquad (3\text{-}73)$$

If $P(T;\ \theta)$ denotes the p.d.f. of the sampling distribution of T, we have

$$P(S) = P(T - \tfrac{1}{2}\ dT \le T \le T + \tfrac{1}{2}\ dT) = P(T;\ \theta)\ dT. \qquad (3\text{-}74)$$

The sets V_1, V_2, and V_3 are mutually exclusive, and the corresponding sets S_1, S_2, and S_3, depending on dT, are mutually exclusive "in probability" to order dT (see Sec. 2.6). Hence

$$P(S) = P(S_1) + P(S_2) + P(S_3). \qquad (3\text{-}75)$$

From Eqs. (3-71) - (3-75) we have

$$P(T;\ \theta) = n\{-f[a(T);\ \theta]\ a'(T) + f[b(T);\ \theta]\ b'(T)\} \left\{\int_{a(T)}^{b(T)} f(x;\ \theta)\,dx\right\}^{n-1}$$

$$(\lambda_1 < T \le \theta,\ n \ge 2). \qquad (3\text{-}76)$$

In virtue of the remark on Lemma 2, the complete sampling distribution of T over the closed interval (λ_1, θ) is given by Eqs. (3-77) and (3-78):

$$P(T;\ \theta) = n\{-f[a(T);\ \theta]\ a'(T) + f[b(T);\ \theta]\ b'(T)\} \left\{\int_{a(T)}^{b(T)} f(x;\ \theta)\ dx\right\}^{n-1}$$

$$[\lambda_1 \leq T \leq \theta \text{ if } P_\theta(\rho) = 0;\ \lambda_1 < T \leq \theta \text{ if } P_\theta(\rho) > 0;\ n \geq 2], \tag{3-77}$$

$$P(T = \lambda_1;\ \theta) = [P_\theta(\rho)]^n \quad [T = \lambda_1 \text{ if } P_\theta(\rho) > 0,\ n \geq 2]. \tag{3-78}$$

It is interesting to note that in Eq. (3-77)

$$P(T;\ \theta) = \frac{d}{dT}\left\{\int_{a(T)}^{b(T)} f(x;\ \theta)\ dx\right\}^{n}. \tag{3-79}$$

The distribution function of the sampling distribution of T can be at once written down with the help of Eq. (3-79). As a useful check, it may also be verified with the help of Eq. (3-79) that the probability measure of the half-open interval $(\lambda_1 < T \leq \theta)$ in Eq. (3-77) is

$$1 - [P_\theta(\rho)]^n. \tag{3-80}$$

Equation (3-80) shows that if $P_\theta(\rho) = 0$, the half-open interval contains the total probability 1, and hence closing the interval by addition of the open endpoint λ_1 does not make any difference. This fact is proved in Lemma 2, where it is shown that if $P_\theta(\rho) = 0$, the point λ_1 is a continuity point (and hence contains zero probability) of the distribution of T. On the other hand, if $P_\theta(\rho) > 0$, Eq. (3-80) shows that the probability measure of the half-open interval falls short of the total probability 1, and that the deficiency $[P_\theta(\rho)]^n$ is made good by addition of the open endpoint λ_1, which then becomes the discrete mass point of the sampling distribution of T.

3.5 The Joint Sampling Distribution of (T, x_i) when $n \geq 2$, Where T is Defined by Equation (3-26), $\theta \geq T \geq b_1^{-1}[x_{(n)}]$

Under the assumptions of Sec. 3.1 and using Eqs. (2-42) - (2-45), the p.d.f. of the joint sampling distribution of $[x_{(1)}, x_{(n)}, x_i]$ when $n \geq 2$ is given by Eqs. (3-81) and (3-83) - (3-85):

$$P[x_{(1)}, x_{(n)}, x_i; \theta] = (n-1)(n-2)\ f[x_{(1)}; \theta]\ f[x_{(n)}; \theta]\ f(x_i; \theta) \left[\int_{x_{(1)}}^{x_{(n)}} f(x; \theta)\ dx\right]^{n-3}$$

$$[a(\theta) \leq x_{(1)} < x_i < x_{(n)} \leq b(\theta),\ n > 2], \qquad (3\text{-}81)$$

where we shall use R_3 to denote the three-dimensional region

$$a(\theta) \leq x_{(1)} < x_i < x_{(n)} \leq b(\theta) \qquad (3\text{-}82)$$

in the space of $[x_{(1)}, x_{(2)}, x_i]$.

$$P[x_{(1)}, x_{(n)}, x_i = x_{(1)}; \theta] = (n-1)\ f[x_{(1)}; \theta]\ f[x_{(n)}; \theta] \left[\int_{x_{(1)}}^{x_{(n)}} f(x; \theta)\ dx\right]^{n-2}$$

$$[a(\theta) \leq x_{(1)} = x_i < x_{(n)} \leq b(\theta),\ n \geq 2], \qquad (3\text{-}83)$$

$$P[x_{(1)}, x_{(n)}, x_i = x_{(n)}; \theta] = (n-1)\ f[x_{(1)}; \theta]\ f[x_{(n)}; \theta] \left[\int_{x_{(1)}}^{x_{(n)}} f(x; \theta)\ dx\right]^{n-2}$$

$$[a(\theta) \leq x_{(1)} < x_i = x_{(n)} \leq b(\theta),\ n \geq 2], \qquad (3\text{-}84)$$

$$P[x_i, x_{(1)} = x_i = x_{(n)}; \theta] = 0$$

$$[a(\theta) \leq x_{(1)} = x_i = x_{(n)} \leq b(\theta),\ n \geq 2]. \qquad (3\text{-}85)$$

We now proceed to obtain the joint sampling distribution of (T, x_i).

First we recall (as mentioned in Sec. 3.4) that the range of possible values of T is the closed interval (λ_1, θ), and that (as shown in Sec. 3.3) the possible values of x_i lie in the interval $[a(T), b(T)]$ for a given T. Hence the region B_{T,x_i} of possible values of (T, x_i) is the set of points (T, x_i) in the two-dimensional space of (T, x_i) such that

$$\lambda_1 \leq T \leq \theta, \quad a(T) \leq x_i \leq b(T). \tag{3-86}$$

We note that since $a(\theta) < b(\theta)$ for all $\theta \in \Omega$, $a(T) < b(T)$ for all $T \neq \lambda_1$. Further $a(T) = b(T)$ if, and only if, ρ is degenerate and $T = \lambda_1$. In virtue of Lemma 2, λ_1 is a discrete mass point of the distribution of T if ρ is nondegenerate and $P_\theta(\rho) > 0$. In view of this, we divide the region R_{T,x_i} into mutually exclusive regions as follows:

Region 1: $\lambda_1 < T \leq \theta$, $\quad a(T) < x_i < b(T)$;

Region 2: $\lambda_1 < T \leq \theta$, $\quad a(T) = x_i < b(T)$;

Region 3: $\lambda_1 < T \leq \theta$, $\quad a(T) < x_i = b(T)$;

Region 4: $T = \lambda_1$, $\quad a(\lambda_1) \leq x_i \leq a(\lambda_2)$ $\quad$ (ρ nondegenerate);

Region 5: $T = \lambda_1$, $\quad a(\lambda_1) = x_i = a(\lambda_2)$ $\quad$ (ρ degenerate).

The joint sampling distribution of (T, x_i) takes different forms in these different regions.

We note the following about Regions 1, 2, and 3: Since we must have either $a(T) = x_{(1)}$ or $b(T) = x_{(n)}$, or $a(T) = x_{(1)}$ and $b(T) = x_{(n)}$; and since $x_i = a(T)$ if, and only if, $x_i = x_{(1)}$ and $x_{(1)} = a(T)$; and since $x_i = b(T)$ if, and only if, $x_i = x_{(n)}$ and $x_{(n)} = b(T)$, it follows that $a(T) < x_i < b(T)$ for all points (T, x_i) in region 1 if, and only if, $x_{(1)} < x_i < x_{(n)}$. Hence to obtain the joint distribution of (T, x_i) in Region 1, we may take as the basis the joint sampling distribution of $[x_{(1)}, x_{(n)}, x_i]$ $[x_{(1)} < x_i < x_{(n)}]$ given in Eq. (3-81). Similarly we may take as the basis the joint sampling distributions (3-83) and (3-84) to obtain the joint distribution of (x_i, T) in Regions 2 and 3 respectively.

REGION 1: $\lambda_1 < T \leq \theta, \quad a(T) < x_i < b(T).$ (3-87)

In this two-dimensional region, the joint sampling distribution of (T, x_i) is nondegenerate and we denote its p.d.f. by $P(T, x_i; \theta)$.

Let S denote the set of points $[x_{(1)}, x_{(n)}, x_i]$ $[x_{(1)} < x_i < x_{(n)}]$ in R_3 (as defined by Eq. (3-82)), which give a value of (T, x_i) in the two-dimensional range

$$(T - \tfrac{1}{2}\,dT,\ T + \tfrac{1}{2}\,dT;\ x_i - \tfrac{1}{2}\,dx_i,\ x_i + \tfrac{1}{2}\,dx_i). \qquad (3\text{-}88)$$

Then, as in the derivation of Eqs. (3-67) - (3-70), we have

$$S = S_1 + S_2 + S_3, \qquad (3\text{-}89)$$

where S_1, S_2, and S_3 are sets of points $[x_{(1)}, x_{(n)}, x_i]$ $[x_{(1)} < x_i < x_{(n)}]$ in R_3 defined as follows:

S_1: $a(T + \tfrac{1}{2}\,dT) \leq x_{(1)} \leq a(T - \tfrac{1}{2}\,dT)$,

$$\ell_1 \leq x_{(n)} < \ell_2, \qquad (3\text{-}90)$$

and $x_i - \tfrac{1}{2}\,dx_i \leq x_i \leq x_i + \tfrac{1}{2}\,dx_i$,

where $a(T + \tfrac{1}{2}\,dT) \leq \ell_1 \leq a(T - \tfrac{1}{2}\,dT)$, and $b(T - \tfrac{1}{2}\,dT) \leq \ell_2 \leq b(T + \tfrac{1}{2}\,dT)$.

S_2: $\ell_1 < x_{(1)} \leq \ell_2$,

$$b(T - \tfrac{1}{2}\,dT) \leq x_{(n)} \leq b(T + \tfrac{1}{2}\,dT), \qquad (3\text{-}91)$$

and $x_i - \tfrac{1}{2}\,dx_i \leq x_i \leq x_i + \tfrac{1}{2}\,dx_i$,

where ℓ_1 and ℓ_2 are as in S_1.

S_3: $a(T + \tfrac{1}{2}\,dT) \leq x_{(1)} \leq a(T - \tfrac{1}{2}\,dT)$,

$$b(T - \tfrac{1}{2}\,dT) \leq x_{(n)} \leq b(T + \tfrac{1}{2}\,dT), \qquad (3\text{-}92)$$

and $x_i - \tfrac{1}{2}\,dx_i \leq x_i \leq x_i + \tfrac{1}{2}\,dx_i$.

Using Eq. (3-81), the probability measures of the sets S_1, S_2, and S_3 to order $dT\,dx_i$ are:

$$P(S_1) = -(n-1)(n-2)\ f[a(T);\ \theta]\ a'(T)\ dT\ f(x_i;\ \theta)\ dx_i$$

$$\times \int_{a(T)}^{b(T)} f[x_{(n)};\ \theta] \left\{\int_{a(T)}^{x_{(n)}} f(x;\ \theta)\ dx\right\}^{n-3} dx_{(n)}$$

$$= -(n-1)(n-2)\ f[a(T);\ \theta]\ a'(T)\ f(x_i;\ \theta)\ dx_i\ dT$$

$$\times \left[\frac{1}{n-2}\left\{\int_{a(T)}^{x_{(n)}} f(x;\ \theta)\ dx\right\}^{n-2}\right]_{x_{(n)}=a(T)}^{x_{(n)}=b(T)}$$

$$= -(n-1)\ f[a(T);\ \theta]\ a'(T)\ f(x_i;\ \theta)\left\{\int_{a(T)}^{b(T)} f(x;\ \theta)\ dx\right\}^{n-2} dT\ dx_i. \tag{3-93}$$

$$P(S_2) = (n-1)(n-2)\ f[b(T);\ \theta]\ b'(T)\ dT\ f(x_i;\ \theta)\ dx_i$$

$$\times \int_{a(T)}^{b(T)} f(x_{(1)};\ \theta) \left\{\int_{x_{(1)}}^{b(T)} f(x;\ \theta)\ dx\right\}^{n-3} dx_{(1)}$$

$$= (n-1)\ f[b(T);\ \theta]\ b'(T)\ f(x_i;\ \theta)\ dT\ dx_i$$

$$\left[-\frac{1}{n-2}\left\{\int_{x_{(1)}}^{b(T)} f(x;\ \theta)\ dx\right\}^{n-2}\right]_{x_{(1)}=a(T)}^{x_{(1)}=b(T)}$$

$$= (n-1)\ f[b(T);\ \theta]\ b'(T)\ f(x_i;\ \theta)\left\{\int_{a(T)}^{b(T)} f(x;\ \theta)\ dx\right\}^{n-2} dT\ dx_i. \tag{3-94}$$

$$P(S_3) = 0 \quad \text{to order } dT\ dx_i. \tag{3-95}$$

Now $P(S) = P(S_1) + P(S_2) + P(S_3)$, (3-96)

to order $dT\ dx_i$, since the sets S_1, S_2, and S_3 are mutually exclusive "in probability" (Sec. 2.6) to order $dT\ dx_i$. Also

$$P(S) = P(T;\ x_i;\ \theta)\ dT\ dx_i. \tag{3-97}$$

Hence Eqs. (3-93) - (3-97) give

$$P(T,\ x_i;\ \theta) = (n-1)\ f(x_i;\ \theta)\ \{-f[a(T);\ \theta]\ a'(T)$$

$$+ f[b(T);\ \theta]\ b'(T)\}\left\{\int_{a(T)}^{b(T)} f(x;\ \theta)\ dx\right\}^{n-2}$$

$$[\lambda_1 < T \le \theta,\ a(T) < x_i < b(T),\ n > 2]. \tag{3-98}$$

REGION 2: $\lambda_1 < T \leq \theta, \quad a(T) = x_i < b(T).$ (3-99)

This is a one-dimensional region in R_{T,x_i}, and the joint distribution of $(T;\ x_i)$ is degenerate in this region. Regarding the distribution of $(T;\ x_i)$ as a one-dimensional distribution in the space of one of the variables, say T, we may denote the p.d.f. of this distribution by $P[T,\ x_i = a(T);\ \theta]$.

$a(T) = x_i < b(T)$ is equivalent to $x_{(1)} = x_i = a(T)$, and $T = a_1^{-1}[x_{(1)}] \geq b_1^{-1}[x_{(n)}]$. $T \geq b_1^{-1}[x_{(n)}]$ gives $b(T) \geq x_{(n)} \geq x_{(1)} = a(T)$. Hence the range of x_n is $a(T) \leq x_{(n)} \leq b(T)$.

Using Eq. (3-83), we have

$$P[T,\ x_i = a(T);\ \theta]$$

$$= -(n-1)\ f[a(T);\ \theta]\ a'(T) \int_{a(T)}^{b(T)} f[x_{(n)};\ \theta] \left(\int_{a(T)}^{x_{(n)}} f(x;\ \theta)\ dx \right)^{n-2} dx_{(n)}$$

$$= -(n-1)\ f[a(T);\ \theta]\ a'(T) \left[\frac{1}{n-1} \left(\int_{a(T)}^{x_{(n)}} f(x;\ \theta)\ dx \right)^{n-1} \right]_{x_{(n)}=a(T)}^{x_{(n)}=b(T)}$$

$$= -f[a(T);\ \theta]\ a'(T) \left(\int_{a(T)}^{b(T)} f(x;\ \theta)\ dx \right)^{n-1}.$$

Hence $P[T,\ x_i = a(T);\ \theta] = -f[a(T);\ \theta]\ a'(T) \left(\int_{a(T)}^{b(T)} f(x;\ \theta)\ dx \right)^{n-1}$

$$[\lambda_1 < T \leq \theta,\ a(T) = x_i < b(T),\ n \geq 2]. \qquad (3\text{-}100)$$

REGION 3: $\lambda_1 < T \leq \theta, \quad a(T) < x_i = b(T).$ (3-101)

The situation here is exactly analogous to that in Region 2, and using Eq. (3-84) we have

$$P[T,\ x_i = b(T);\ \theta] = f[b(T);\ \theta]\ b'(T) \left(\int_{a(T)}^{b(T)} f(x;\ \theta)\ dx \right)^{n-1}$$

$$[\lambda_1 < T \leq \theta,\ a(T) < x_i = b(T),\ n \geq 2]. \qquad (3\text{-}102)$$

REGION 4: $T = \lambda_1, \quad a(\lambda_1) \le x_i \le b(\lambda_1)$ (ρ nondegenerate). (3-103)

In virtue of Lemma 1 of Sec. 3.4, $T = \lambda_1$ if, and only if, $x_{(1)} \in \rho$ and $x_{(n)} \in \rho$. Let $P(T = \lambda_1, x_i; \theta)$ denote the p.d.f. of the distribution of x_i for $T = \lambda_1$. Then $P(T = \lambda_1, x_i; \theta)\, dx_i$ is the probability that a random value of x_i falls in the range $(x_i - \frac{1}{2} dx_i, x_i + \frac{1}{2} dx_i)$ belonging to ρ, and all the remaining $n-1$ sample members fall in the interval $\rho = [a(\lambda_1), b(\lambda_1)]$. Hence

$$P(T = \lambda_1, x_i; \theta) = f(x_i; \theta) \left\{ \int_{a(\lambda_1)}^{a(\lambda_2)} f(x; \theta)\, dx \right\}^{n-1} = f(x_i; \theta)\, [P_\theta(\rho)]^{n-1}$$

$$[T = \lambda_1,\ a(\lambda_1) \le x_i \le b(\lambda_1),\ n \ge 2]. \qquad (3\text{-}104)$$

REGION 5: $T = \lambda_1, \quad a(\lambda_1) = x_i = b(\lambda_1)$ (ρ degenerate). (3-105)

In virtue of Lemma 2 of Sec. 3.4, $P(T = \lambda_1) = 0$. We have

$$P[T = \lambda_1,\ a(\lambda_1) = x_i = b(\lambda_1);\ \theta] = 0.$$

$$[T = \lambda_1,\ a(\lambda_1) = x_i = b(\lambda_1),\ n \ge 2]. \qquad (3\text{-}106)$$

Thus Eqs. (3-98), (3-100), (3-102), (3-104), and (3-106) give the joint distribution of (T, x_i) when $n \ge 2$.

As a useful check on the results of this section, it may be noted that the sampling distribution of T in Sec. 3.4 is deducible by integration from the joint sampling distribution of (T, x_i) in this section.

3.6 The Conditional Distribution of x_i for a Given T, when $n \ge 2$, Where T is Defined by Equation (3-26), $\theta \ge T \ge b_1^{-1}[x_{(n)}]$

The conditional distribution of x_i for a given T when $n \ge 2$ can now be written down from the joint distribution of (x_i, T) of Sec. 3.5 and the marginal distribution of T of Sec. 3.4.

We make the following cases:

CASE I: $\lambda_1 < T \leq \theta$, $a(T) < x_i < b(T)$, $n > 2$.

Here the conditional distribution of x_i for a given T is nondegenerate and possesses a p.d.f. $P(x_i; \theta|T)$ given by

$$P(x_i; \theta|T) = \frac{P(T, x_i; \theta)}{P(T; \theta)} . \tag{3-107}$$

Using Eqs. (3-77) and (3-98) in Eq. (3-107), we have

$$P(x_i; \theta|T) = \frac{n-1}{n} \cdot \frac{f(x_i; \theta)}{\int_{a(T)}^{b(T)} f(x; \theta)\, dx}$$

$$[\lambda_1 < T \leq \theta,\ a(T) < x_i < b(T),\ n > 2]. \tag{3-108}$$

It should be noted that by the definition of a conditional distribution, Eq. (3-107) is defined at every point T at which $P(T; \theta) > 0$.

CASE II: $\lambda_1 < T \leq \theta$, $a(T) = x_i < b(T)$, $n \geq 2$.

Here the point $x_i = a(T)$ is a discrete mass point of the conditional distribution of x_i for a given T, and we use the symbol $P[x_i = a(T); \theta|T]$ to denote the conditional probability at the point $x_i = a(T)$.

We have

$$P[x_i = a(T); \theta|T] = \frac{P[T, x_i = a(T); \theta]}{P(T; \theta)} . \tag{3-109}$$

Using Eqs. (3-77) and (3-100) in Eq. (3-109), we have

$$P[x_i = a(T); \theta|T] = \frac{1}{n} \frac{-f[a(T); \theta]\, a'(T)}{-f[a(T); \theta]\, a'(T) + f[b(T); \theta]\, b'(T)}$$

$$[\lambda_1 < T \leq \theta,\ a(T) = x_i < b(T),\ n \geq 2]. \tag{3-110}$$

CASE III. $\lambda_1 < T \leq \theta$, $a(T) < x_i = b(T)$, $n \geq 2$.

The point $x_i = b(T)$ is a discrete mass point of the conditional distribution of x_i for a given T. Equations (3-77) and (3-102) give

$$P[x_i = b(T);\ \theta | T] = \frac{1}{n} \frac{f[b(T);\ \theta]\ b'(T)}{-f[a(T);\ \theta]\ a'(T) + f[b(T);\ \theta]\ b'(T)}$$

$$[\lambda_1 < T \leq \theta,\quad a(T) < x_i = b(T),\quad n \geq 2]. \qquad (3\text{-}111)$$

CASE IV: $T = \lambda_1$, $a(\lambda_1) \leq x_i \leq b(\lambda_1)$, $n \geq 2$, nondegenerate, and $P_\theta(\rho) > 0$.

Here $T = \lambda_1$ is a discrete mass point of the distribution of T, and if $P(x_i;\ \theta | T = \lambda_1)$ denotes the p.d.f. of the conditional distribution of x_i, we have

$$P(x_i;\ \theta | T = \lambda_1) = \frac{P(T = \lambda_1,\ x_i;\ \theta)}{P(T = \lambda_1)}. \qquad (3\text{-}112)$$

Using Eqs. (3-78) and (3-104) in Eq. (3-112), we have

$$P(x_i;\ \theta | T = \lambda_1) = \frac{f(x_i;\ \theta)}{P_\theta(\rho)}$$

$$[T = \lambda_1,\ a(\lambda_1) \leq x_i \leq b(\lambda_1),\ n \geq 2]. \qquad (3\text{-}113)$$

Thus Eqs. (3-108), (3-110), (3-111), and (3-113) give the conditional distribution of x_i for a given T. It may be noted that for every fixed T in $(\lambda_1 < T \leq \theta)$, the probability measure of the two discrete mass points a(T) and b(T) of the conditional distribution of x_i is 1/n, and the probability measure of the open interval [a(T), b(T)] is (n - 1)/n. If $T = \lambda_1$ and $P_\theta(\rho) > 0$, the total probability of the conditional distribution of x_i is concentrated in the interval ρ.

3.7 The General Form of $f(x;\ \theta)$ for Existence of a Single Sufficient Statistic for θ when $n \geq 2$ Under the Assumptions of Section 3.1

CASE I: $a(\theta)$ and $b(\theta)$ are monotonic Type I.

THEOREM 2

Under the assumptions of Sec. 3.1, where $a(\theta)$ and $b(\theta)$ are monotonic Type I, a necessary and sufficient condition that the single statistic T as defined by Eq. (3-26) is sufficient for θ, for all random samples of any size $n \geq 2$ and for all $\theta \in \Omega$, is that for all $x \in R$ and $\theta \in \Omega$, $f(x;\ \theta)$ is of the form

$$f(x;\ \theta) = g(x)\ h(\theta), \tag{3-114}$$

where $g(x)$ is a function of x defined in R, and $h(\theta)$ is a function of θ defined in Ω.

PROOF. By the equivalent Form A of the definition of a single sufficient statistic (see Sec. 2.1), the statistic T, as defined by Eq. (3-26), is sufficient for θ for all random samples of any size $n \geq 2$ and for all $\theta \in \Omega$ if, and only if, the conditional distribution of x_i for every fixed T, as given by Eqs. (3-108), (3-110), (3-111), and (3-113), is independent of θ for all $x_i \in R$, $\theta \in \Omega$, and $n \geq 2$.

In Eq. (3-108), $P(x_i;\ \theta|T)$ is independent of θ for every fixed T in $(\lambda_1 < T \leq \theta)$ for all $x_i \in R$, $\theta \in \Omega$, and $n > 2$, if, and only if, $f(x_i;\ \theta)/f(x_j;\ \theta)$ is independent of θ for all x_i, $x_j \in R$ and $\theta \in \Omega$, that is (in virtue of Sec. 2.4), if, and only if, $f(x;\ \theta)$ is of the form (3-114) for all $x \in R$ and $\theta \in \Omega$.

In Eq. (3-110), $P(x_i = a(T);\ \theta|T)$ is independent of θ for every fixed T in $(\lambda_1 < T \leq \theta)$, all $x_i = a(T) \in R$, $\theta \in \Omega$, and $n \geq 2$, if and only if, $f[b(T);\ \theta]/f[a(T);\ \theta]$ is independent of θ for all $a(T)$, $b(T)$, $\in R$, and $\theta \in \Omega$, that is (in virtue of Sec. 2.5), if, and only if, $f(x;\ \theta)$ is of the form (3-114) for all $x \in R$ and $\theta \in \Omega$.

Similarly, in Eq. (3-111), $P[x_i = b(T);\ \theta|T]$ is independent of θ for every fixed T in $(\lambda_1 < T \leq \theta)$, all $x_i = b(T) \in R$, $\theta \in \Omega$, and $n \geq 2$, if, and only if, $f(x;\ \theta)$ is of the form (3-114) for all $x \in R$ and $\theta \in \Omega$.

In Eq. (3-113), $P(x_i;\ \theta|T = \lambda_1)$ is independent of θ for $T = \lambda_1$, all $x_i \in \rho \subset R$, $\theta \in \Omega$, and $n \geq 2$, if, and only if, $f(x_i;\ \theta)/f(x_j;\ \theta)$ is independent of θ for all $x_i, x_j \in \rho \subset R$, and $\theta \in \Omega$, that is (in virtue of Sec. 2.5), if, and only if, $f(x;\ \theta)$ is of the form (3-114) for all $x \in \rho \subset R$ and $\theta \in \Omega$.

Thus all cases of the conditional distribution of x_i for every fixed T when $n \geq 2$ give the same form (3-114) for $f(x; \theta)$, as a necessary and sufficient condition for the single statistic T to be sufficient for the parameter θ. Hence the theorem is proved.

The results of Theorem 1 (in Sec. 3.3) and Theorem 2 (in the present section) can now be combined into Theorem 3.

THEOREM 3

Under the assumptions of Sec. 3.1, where $a(\theta)$ and $b(\theta)$ are monotonic Type I, a necessary and sufficient condition for the existence of a single sufficient statistic for θ for all random samples of any size $n \geq 2$ and for all $\theta \in \Omega$ is that for all $x \in R$ and $\theta \in \Omega$, $f(x; \theta)$ is of the form (3-114), in which case the single statistic T, defined by Eq. (3-26), is sufficient for θ.

CASE II: $a(\theta)$ and $b(\theta)$ are monotonic Type II.

In Secs. 3.3 - 3.6 the results were obtained for the monotonic Type I case. Exactly analogous results hold good for the monotonic Type II case. We shall now quote some of the results, omitting proofs.

We define in this case the single statistic T by

$$T = \min\{a_2^{-1}[x_{(1)}],\ b_2^{-1}[x_{(n)}]\}. \tag{3-115}$$

The statistic T has the property that when T is fixed, the range of each x_i lies in the range $[a(T), b(T)]$, which is independent of θ for all $\theta \in \Omega$ for all sample points $(x_1, \ldots, x_n)$ $(n \geq 2)$. Theorem 1 (Sec. 3.3) remains true for the monotonic Type II case with T defined by Eq. (3-115). Lemmas 1 and 2 of Sec. 3.4 remain true if λ_1 is replaced by λ_2, and ρ is taken as $[a(\lambda_2), b(\lambda_2)]$. The range of the sampling distribution of T is the closed interval $(\theta \leq T < \lambda_2)$. As in Eqs. (3-77) and (3-78), the sampling distribution of T as defined by Eq. (3-115) for the monotonic Type II case is given by Eqs. (3-116) and (3-117):

$$P(T;\ \theta) = n\{f[a(T);\ \theta]\ a'(T) - f[b(T);\ \theta]\ b'(T)\} \left[\int_{a(T)}^{b(T)} f(x;\ \theta)\ dx\right]^{n-1}$$

$$[\theta \leq T \leq \lambda_2 \text{ if } P_\theta(\rho) = 0;\ \theta \leq T < \lambda_2 \text{ if } P_\theta(\rho) > 0;\ n \geq 2], \quad (3\text{-}116)$$

and $P(T = \lambda_2;\ \theta) = [P_\theta(\rho)]^n$

$$[T = \lambda_2 \text{ if } P_\theta(\rho) > 0;\ n \geq 2]. \quad (3\text{-}117)$$

It should be remembered that in Eq. (3-116), $a'(T) \geq 0$ and $b'(T) \leq 0$, since $a(\theta)$ is m.i. and $b(\theta)$ is m.d.

The joint sampling distribution of (T, x_i) for the monotonic Type II case is obtained from Eqs. (3-98), (3-100), (3-102), (3-104), and (3-106) by writing λ_2 for λ_1, (θ, λ_2) for the interval (λ_1, θ), $-a'(T)$ for $a'(T)$, and $-b'(T)$ for $b'(T)$. The same substitutions in Eqs. (3-108), (3-110), (3-111), and (3-113) yield the conditional distribution of x_i for a fixed T for the monotonic Type II case.

Theorems 2 and 3 remain true for the monotonic Type II case, where T is defined by Eq. (3-115).

Combining the results of monotonic Type I and Type II cases, Theorem 3 may now be extended in the following form:

THEOREM 4

Under the assumptions of Sec. 3.1, a necessary and sufficient condition for the existence of a single sufficient statistic for θ for all random samples of any size $n \geq 2$ and for all $\theta \in \Omega$ is that for all $x \in R$ and $\theta \in \Omega$, $f(x;\ \theta)$ is of the form (3-114), in which case the single sufficient statistic T is defined by Eq. (3-26) or Eq. (3-115) according as $a(\theta)$ and $b(\theta)$ are respectively monotonic Type I or monotonic Type II.

3.8 The Sampling Distribution of a Single Sufficient Statistic in the Nonregular Case of a Single Unknown Parameter

Under the assumptions of Sec. 3.1, Eqs. (3-77) and (3-78) give the sampling distribution of T, as defined by Eq. (3-26), for the monotonic Type I case; and Eqs. (3-116) and (3-117) give the sampling distribution of T, as defined by Eq. (3-115), for the monotonic Type II case.

If we put

$$f(x;\ \theta) = g(x)\ h(\theta), \tag{3-118}$$

T becomes a single sufficient statistic for θ, and the sampling distribution of the sufficient statistic T is then given by the substitution (3-118), in Eqs. (3-77) and (3-78), and in Eqs. (3-116) and (3-117).

First we note that since

$$\int_{a(\theta)}^{b(\theta)} f(x;\ \theta)\ dx = 1 \qquad \text{for all } \theta \in \Omega, \tag{3-119}$$

Eq. (3-118) gives

$$\int_{a(\theta)}^{b(\theta)} g(x)\ dx = \frac{1}{h(\theta)} \qquad \text{for all } \theta \in \Omega. \tag{3-120}$$

It follows from Eq. (3-120) that $h(\theta)$ is m.d. or m.i. according as $a(\theta)$ and $b(\theta)$ are monotonic Type I or monotonic Type II. Let $G(x) = \int g(x)\ dx$ [i.e. primitive of $g(x)$], which exists for all $x \in R$, since the given family of distributions possesses p.d.f. $g(x)\ h(\theta)$ for all $x \in R$ and $\theta \in \Omega$. Then Eq. (3-120) becomes

$$G[b(\theta)] - G[a(\theta)] = \frac{1}{h(\theta)} \qquad \text{for all } \theta \in \Omega. \tag{3-121}$$

Since $G(x)$ is a differentiable function of x for all $x \in R$, and $a(\theta)$ and $b(\theta)$ are differentiable functions of θ for all $\theta \in \Omega$, it follows that the left-hand member of Eq. (3-121) is a differentiable function of θ for all $\theta \in \Omega$. Hence $h(\theta)$ is a differentiable function of θ for all $\theta \in \Omega$. Differentiating Eq. (3-121) with respect to θ, we have, for all $\theta \in \Omega$,

$$g[b(\theta)]\ b'(\theta) - g[a(\theta)]\ a'(\theta) = -\frac{h'(\theta)}{[h(\theta)]^2}. \tag{3-122}$$

The substitution (3-118) gives

$$P_\theta(\rho) = \int_{(\rho)} f(x;\ \theta)\ dx = h(\theta) \int_{(\rho)} g(x)\ dx. \tag{3-123}$$

For the monotonic Type I case, substitution (3-118) in Eq. (3-77) gives

$$P(T;\ \theta) = n[h(\theta)]^n\{-g[a(T)]\ a'(T) + g[b(T)]\ b'(T)\} \left(\int_{a(T)}^{b(T)} g(x)\ dx \right)^{n-1}$$

$$[\lambda_1 \leq T \leq \theta \text{ if } P_\theta(\rho) = 0;\ \lambda_1 < T \leq \theta \text{ if } P_\theta(\rho) > 0;\ n \geq 2]. \tag{3-124}$$

In Eq. (3-124) every value of T belongs to Ω. Hence Eqs. (3-120) and (3-122) give

$$\int_{a(T)}^{b(T)} g(x)\ dx = \frac{1}{h(T)} \tag{3-125}$$

and $$-g[a(T)]\ a'(T) + g[b(T)]\ b'(T) = -\frac{h'(T)}{[h(T)]^2}. \tag{3-126}$$

Using Eqs. (3-125), (3-126), Eq. (3-124) becomes

$$P(T;\ \theta) = -\frac{n[h(\theta)]^n\ h'(T)}{[h(T)]^{n+1}}$$

$$[\lambda_1 \leq T \leq \theta \text{ if } P_\theta(\rho) = 0;\ \lambda_1 < T \leq \theta \text{ if } P_\theta(\rho) > 0;\ n \geq 2]. \tag{3-127}$$

Also, substitution of Eq. (3-123) in Eq. (3-78) gives

$$P(T = \lambda_1;\ \theta) = [h(\theta)]^n \left(\int_{a(\lambda_1)}^{b(\lambda_1)} g(x)\ dx \right)^n$$

$$[T = \lambda_1 \text{ if } P_\theta(\rho) > 0;\ n \geq 2]. \tag{3-128}$$

Equations (3-127) and (3-128) give the sampling distribution of the sufficient statistic T for the monotonic Type I case. Similarly, for the monotonic Type II case, the sampling distribution of the sufficient statistic T is given by Eqs. (3-129) and (3-130):

$$P(T;\ \theta) = \frac{n[h(\theta)]^n\ h'(T)}{[h(T)]^{n+1}}$$

$$[\theta \leq T \leq \lambda_2 \text{ if } P_\theta(\rho) = 0;\ \theta \leq T < \lambda_2 \text{ if } P_\theta(\rho) > 0;\ n \geq 2], \tag{3-129}$$

$$\text{and} \quad P(T = \lambda_2;\ \theta) = [h(\theta)]^n \left\{\int_{a(\lambda_2)}^{b(\lambda_2)} g(x)\ dx\right\}^n$$

$$[T = \lambda_2 \text{ if } P_\theta(\rho) > 0;\ n \geq 2]. \qquad (3\text{-}130)$$

The sampling distribution of the sufficient statistic T for the particular case when $P_\theta(\rho) = 0$ was given by the author (Huzurbazar, 1955) and used by him in obtaining confidence intervals for θ. But the results in this section have been obtained more rigorously besides being of a more general nature, than those previously given by him.

3.9 Non-existence of a Single Sufficient Statistic when a(θ) and b(θ) are Monotonic in θ, but not in Opposite Senses

We now replace the assumptions 1, 2, and 3 of Sec. 3.1 by the following assumptions:

1. For all $\theta \in \Omega$, $a(\theta)$ and $b(\theta)$ are continuous and monotonic in θ, but not in opposite senses.
2. For all $\theta \in \Omega$, $a(\theta) < b(\theta)$.

As already mentioned in the first paragraph of Sec. 3.1, at least one of the functions $a(\theta)$ and $b(\theta)$ is not constant for all $\theta \in \Omega$, the case being nonregular. It may be noted that assumption 1 implies that none of $a(\theta)$ and $b(\theta)$ is constant for all $\theta \in \Omega$, since otherwise they may be regarded as being monotonic in opposite senses. It is for this reason that we prefer to state assumption 1 in the form given above rather than state it in the form: "$a(\theta)$ and $b(\theta)$ are monotonic in the same sense," which would include the possibility that one of $a(\theta)$ and $b(\theta)$ is constant for all $\theta \in \Omega$.

THEOREM 5

Under assumptions 1 and 2 of this section, there exists no single statistic T with the property that, for every fixed value of T, the range of each x_i $(i = 1, \ldots, n)$ is independent of θ for all $\theta \in \Omega$ and for all random samples of any size $n \geq 2$.

PROOF. First suppose that $a(\theta)$ is m.i. and $b(\theta)$ is m.i. If the theorem is not true, there exists a single statistic T with the property mentioned in the statement of the theorem. Then we must have, for all T, all random samples of any size $n \geq 2$, and all $\theta \in \Omega$,

$$a(\theta) \leq \phi_1(T) \leq x_{(1)} \leq x_{(n)} \leq \phi_2(T) \leq b(\theta), \tag{3-131}$$

where $\phi_1(T)$ and $\phi_2(T)$ are functions of T.

Let I_a and I_b be respectively the intervals of values of $a(\theta)$ and $b(\theta)$ for all $\theta \in \Omega$. Let $x_{(1)} \in I_a$ and $x_{(n)} \in I_b$. It follows that $\phi_1(T) \in I_a$ and $\phi_2(T) \in I_b$.

Let θ_0 be the value of θ for the particular member of the family of distributions from which we are sampling. Then

$$a(\theta_0) \leq \phi_1(T) \leq x_{(1)} \leq x_{(n)} \leq \phi_2(T) \leq b(\theta_0). \tag{3-132}$$

From Eq. (3-132), it follows that if $x_{(1)} = a(\theta_0)$, then

$$\phi_1(T) = a(\theta_0). \tag{3-133}$$

Solving Eq. (3-133) for θ_0, we have

$$\theta_o = \text{any point of the closed interval } a_1^{-1}[\phi_1(T)], a_2^{-1}[\phi_1(T)]. \tag{3-134}$$

From Eq. (3-132), if $x_{(n)} = b(\theta_0)$, then

$$\phi_2(T) = b(\theta_0), \tag{3-135}$$

which gives

$$\theta_0 = \text{any point of the closed interval } b_1^{-1}[\phi_2(T)], b_2^{-1}[\phi_2(T)]. \tag{3-136}$$

Since the two equations $x_{(1)} = a(\theta_0)$ and $x_{(n)} = b(\theta_0)$ must give the same set of values of θ_0, the two intervals in Eqs. (3-134) and (3-136) must be the same. Hence

$$a_1^{-1}[\phi_1(T)] = b_1^{-1}[\phi_2(T)], \tag{3-137}$$

and $a_2^{-1}[\phi_1(T)] = b_2^{-1}[\phi_2(T)]$. (3-138)

Now consider Eq. (3-131).

The statement $a(\theta) \leq \phi_1(T)$ is equivalent to

$$\theta \leq a_2^{-1}[\phi_1(T)], \tag{3-139}$$

and the statement $\phi_2(T) \leq b(\theta)$ is equivalent to

$$b_1^{-1}[\phi_2(T)] \leq \theta. \tag{3-140}$$

Hence Eqs. (3-137) - (3-140) give

$$a_1^{-1}[\phi_1(T)] = b_1^{-1}[\phi_2(T)] \leq \theta \leq a_2^{-1}[\phi_1(T)] = b_2^{-1}[\phi_2(T)]. \tag{3-141}$$

Now $\phi_1(T) \in I_a$ and $\phi_2(T) \in I_b$. As remarked in Sec. 2.3, we have, for almost all $\phi_1(T) \in I_a$,

$$a_1^{-1}[\phi_1(T)] = a_2^{-1}[\phi_1(T)]. \tag{3-142}$$

Then Eqs. (3-141) and (3-142) give

$$a_1^{-1}[\phi_1(T)] = b_1^{-1}[\phi_2(T)] = \theta = a_2^{-1}[\phi_1(T)] = b_2^{-1}[\phi_2(T)] \tag{3-143}$$

for almost all $\phi_1(T) \in I_a$ and all $\phi_2(T) \in I_b$. Equation (3-143) implies

$$a(\theta) = \phi_1(T) \quad \text{and} \quad b(\theta) = \phi_2(T) \tag{3-144}$$

for almost all $\phi_1(T) \in I_a$ and all $\phi_2(T) \in I_b$. Equation (3-144) is absurd, since $\phi_1(T)$ and $\phi_2(T)$ are independent of θ, being functions of the statistic T; and none of $a(\theta)$ and $b(\theta)$ is constant for all $\theta \in \Omega$.

Similarly, if $a(\theta)$ is m.d. and $b(\theta)$ is m.d., there exists no single statistic T with the property mentioned in the statement of the theorem.

Since a single sufficient statistic, if it exists, must possess the property referred to in Theorem 5, we have the following theorem.

THEOREM 6 (Pitman's conjecture)

Under assumptions 1 and 2 of this section, there exists no single statistic which is sufficient for θ for all $\theta \in \Omega$ for all random samples of any size $n \geq 2$.

COROLLARY. If, for all $\theta \in \Omega$, $a(\theta)$ and $b(\theta)$ are continuous, monotonic functions of θ, and $a(\theta) < b(\theta)$, then a necessary condition for the existence of a single sufficient statistic for all $\theta \in \Omega$ and for all random samples of any size $n \geq 2$ is that $a(\theta)$ and $b(\theta)$ are monotonic in θ in opposite senses.

3.10 The General Form of Distributions of the Nonregular Type, Depending on a Single Unknown Parameter θ and Admitting a Single Sufficient Statistic for θ

The results of Theorems 4 and 6 can be combined into the central theorem of this discussion.

THEOREM 7

ASSUMPTIONS. Let $f(x; \theta)$ be the p.d.f. of a family of distributions of the nonregular type, depending on a single unknown parameter θ, and with ranges $[a(\theta), b(\theta)]$, where the range of admissible values of θ is some nondegenerate interval $\Omega = (\lambda_1, \lambda_2)$ which may be closed, open, half-open, finite, or infinite. Let the set R be the union of the ranges $[a(\theta), b(\theta)]$ for all $\theta \in \Omega$. For all $\theta \in \Omega$, $a(\theta)$ and $b(\theta)$ are continuous, differentiable, monotonic function of θ, and $a(\theta) < b(\theta)$.

CONCLUSION. For the existence of a single sufficient statistic for θ for all $\theta \in \Omega$ for all random samples of any size $n \geq 2$, it is both necessary and sufficient that (1) $a(\theta)$ and $b(\theta)$ are monotonic in θ in opposite senses, and that (2) for all $x \in R$ and all $\theta \in \Omega$, $f(x; \theta)$ is of the form

$$f(x; \theta) = g(x)\, h(\theta). \tag{3-145}$$

Further, when conditions (1) and (2) are satisfied, the single statistic T defined by

$$T = \max\{a_1^{-1}[x_{(1)}],\ b_1^{-1}[x_{(n)}]\} \tag{3-146}$$

is sufficient for θ when $a(\theta)$ and $b(\theta)$ are monotonic Type I; and the single statistic T defined by

$$T = \min\{a_2^{-1}[x_{(1)}],\ b_2^{-1}[x_{(n)}]\} \tag{3-147}$$

is sufficient for θ when $a(\theta)$ and $b(\theta)$ are monotonic Type II.

The sampling distribution of the sufficient statistic T is given by Eqs. (3-127) and (3-128) for the monotonic Type I case, and by Eqs. (3-129) and (3-130) for the monotonic Type II case.

Finally it may be remarked that since the single sufficient statistic T is a function of only the smallest and the greatest members in the sample, the nonregular case investigated in this section belongs to the pure type (see Sec. 2.10). It may also be noted that apart from the two extreme observations in the sample, only the terminals a and b, and not the p.d.f., enter into the definition of the single sufficient statistic T.

3.11 The Particular Case when Only One Terminal of the Range Depends on θ

If only one terminal of the range, say $a(\theta)$, depends on θ and is a continuous, differentiable, monotonic function of θ for all $\theta \in \Omega$, and the other terminal b is independent of θ, then we can regard b to be monotonic in θ in the sense opposite to that of $a(\theta)$. Thus the results obtained previously hold good in this particular case also. In this case, I_b consists of the single point $b(\lambda_1) = b(\lambda_2)$, which belongs to ρ also according to our convention. Hence Eq. (3-17) becomes

$$R = I_a + \rho. \tag{3-148}$$

Here, for all $x_{(n)} \in R$,

$$b_1^{-1}[x_{(n)}] = \lambda_1, \quad b_2^{-1}[x_{(n)}] = \lambda_2, \tag{3-149}$$

If $a(\theta)$ is m.d., Eq. (3-146) becomes

$$T = \max\{a_1^{-1}[x_{(1)}], \lambda_1\} = a_1^{-1}[x_{(1)}]. \tag{3-150}$$

If $a(\theta)$ is m.i., Eq. (3-147) gives

$$T = \min\{a_2^{-1}[x_{(1)}], \lambda_2\} = a_2^{-1}[x_{(1)}]. \tag{3-151}$$

Similarly, if the lower terminal a is independent of θ but $b(\theta)$ depends on θ, we have

$$T = b_1^{-1}[x_{(n)}] \tag{3-152}$$

if $b(\theta)$ is m.i., and

$$T = b_2^{-1}[x_{(n)}] \tag{3-153}$$

if $b(\theta)$ is m.d.

3.12 Illustrative Examples

We consider a few examples to illustrate the notation, methods used, and some results obtained in the previous sections.

EXAMPLE 1. Consider the rectangular distribution with p.d.f.

$$f(x;\ \theta) = \frac{1}{5\theta} \quad (-2\theta \leq x \leq 3\theta,\ \theta > 0).$$

Here $\Omega = (\lambda_1, \lambda_2)$ is the open infinite interval $(0, \infty)$, so that $\lambda_1 = 0$ and $\lambda_2 = \infty$. $a(\theta) = -2\theta$ and $b(\theta) = 3\theta$ are monotonic Type I and differentiable in Ω.

$$g(x) = 1, \quad h(\theta) = \frac{1}{5(\theta)} .$$

I_a is the open infinite interval $(-\infty, 0)$, and I_b is the open infinite interval $(0, \infty)$. ρ consists of the single point 0, and $R = (-\infty, \infty)$.

$$\text{If } x_{(1)} \leq 0, \quad a_1^{-1}[x_{(1)}] = -\frac{1}{2}x_{(1)},$$

$$\text{and if } x_{(1)} \geq 0, \quad a_1^{-1}[x_{(1)}] = \lambda_1 = 0.$$

If $x_{(n)} \geq 0$, $b_1^{-1}[x_{(n)}] = \frac{1}{3} x_{(n)}$,

and if $x_{(n)} \leq 0$, $b_1^{-1}[x_{(n)}] = \lambda_1 = 0$.

In virtue of Theorem 7, the single statistic T is sufficient for θ, where

$$T = \max\{a_1^{-1}[x_{(1)}], b_1^{-1}[x_{(n)}]\}$$
$$= \max\{-\frac{1}{2} x_{(1)}, \frac{1}{3} x_{(n)}\}.$$

Since ρ is degenerate, the sampling distribution of T has no discrete mass point and is given by substitution in Eq. (3-127):

$$P(T;\ \theta) = -\frac{n(1/5\theta)^n\ (-1/5T^2)}{(1/5T)^{n+1}} \qquad (0 \leq T \leq \theta)$$

$$= \frac{n\ T^{n-1}}{\theta^n} \qquad (0 \leq T \leq \theta)$$

EXAMPLE 2. Consider the distribution

$$f(x;\ \theta) = \frac{2x}{4 + 4\cos\theta + \cos 2\theta}$$
$$[\sin\theta \leq x \leq 2 + \cos\theta;\ 0 \leq \theta \leq \pi/2].$$

Here $g(x) = 2x$, $h(\theta) = \dfrac{1}{4 + 4\cos\theta + \cos 2\theta}$.

$\Omega = (\lambda_1, \lambda_2)$ is the closed interval $(0, \pi/2)$, with $\lambda_1 = 0$ and $\lambda_2 = \pi/2$. $a(\theta) = \sin\theta$ and $b(\theta) = 2 + \cos\theta$ are monotonic Type II and differentiable in Ω. I_a is the closed interval (0, 1), I_b is the closed interval (2, 3). ρ is the closed interval (1, 2), R is the closed interval (0, 3). This is the case in which ρ is non-degenerate. In virtue of Theorem 7, the single statistic T is sufficient for θ, where

$$T = \min\{a_2^{-1}[x_{(1)}], b_2^{-1}[x_{(n)}]\}.$$

We note that

$$a_2^{-1}[x_{(1)}] = \sin^{-1}[x_{(1)}] \qquad \text{if } 0 \le x_{(1)} \le 1,$$

$$= \frac{\pi}{2} \qquad \text{if } 1 \le x_{(1)} \le \frac{\pi}{2}.$$

$$b_2^{-1}[x_{(n)}] = \cos^{-1}[x_{(n)} -2] \qquad \text{if } 2 \le x_{(n)} \le 3,$$

$$= \frac{\pi}{2} \qquad \text{if } 0 \le x_{(n)} \le 2.$$

Hence

$$T = \sin^{-1}[x_{(1)}] \qquad \text{if } 0 \le x_{(1)} \le 1,\ 0 \le x_{(n)} \le 2;$$

$$= \min\{\sin^{-1}[x_{(1)}],\ \cos^{-1}[x_{(n)} - 2]\} \text{ if } 0 \le x_{(1)} \le 1,\ 2 \le x_{(n)} \le 3;$$

$$= \cos^{-1}[x_{(n)} - 2] \qquad \text{if } 1 \le x_{(1)} \le 3,\ 2 \le x_{(n)} \le 3;$$

$$= \frac{\pi}{2} \qquad \text{if } 1 \le x_{(1)} \le x_{(n)} \le 2.$$

The sampling distribution of T has a discrete mass point at $T = \pi/2$. Substitution in Eqs. (3-129) and (3-130) gives the sampling distribution of T:

$$P(T;\ \theta) = \frac{2n(2 \sin T + \sin 2T)\ (4 + 4 \cos T + \cos 2T)^{n-1}}{(4 + 4 \cos \theta + \cos 2\theta)^n} \qquad (\theta \le T < \pi/2),$$

and $$P(T = \frac{\pi}{2};\ \theta) = \frac{3^n}{(4 + 4 \cos \theta + \cos 2\theta)^n}$$

EXAMPLE 3. Consider the rectangular distribution

$$f(x;\ \theta) = \frac{1}{b(\theta) - a(\theta)} \qquad [a(\theta) \le x \le b(\theta)],$$

where $a(\theta)$ and $b(\theta)$ are defined in the closed interval (0, 3) by

$$a(\theta) = 0 \qquad \text{if } 0 \le \theta \le 1,$$

$$= (\theta - 1)^2 \qquad \text{if } 1 \le \theta \le 3.$$

$$b(\theta) = (2 - \theta)^3 + 5 \qquad \text{if } 0 \le \theta \le 2,$$

$$= 5, \qquad \text{if } 2 \le \theta \le 3.$$

It will be seen that $a(\theta)$ and $b(\theta)$ are monotonic Type II, continuous, and differentiable in $\Omega = (0, 3)$. $\lambda_1 = 0$, $\lambda_2 = 3$. I_a is the closed interval (0, 4), I_b is the closed interval (5, 13). ρ is the closed interval (4, 5), R is the closed interval (0, 13).

$$g(x) = 1, \quad h(\theta) = \frac{1}{b(\theta) - a(\theta)},$$

where $a(\theta)$ and $b(\theta)$ are as already defined. It may be noted that

$$\begin{aligned} f(x; \theta) = h(\theta) &= \frac{1}{(2 - \theta)^3 + 5} && \text{if } 0 \le \theta \le 1, \\ &= \frac{1}{(2 - \theta)^3 + 5 - (\theta - 1)^2} && \text{if } 1 \le \theta \le 2, \\ &= \frac{1}{5 - (\theta - 1)^2} && \text{if } 2 \le \theta \le 3. \end{aligned}$$

$a(\theta)$ is constant in the closed interval (0, 1), and $b(\theta)$ is constant in the closed interval (2, 3). Theorem 7 gives the single statistic T sufficient for θ where

$$T = \min\{a_1^{-1}[x_{(1)}], b_2^{-1}[x_{(n)}]\}.$$

We note that

$$\begin{aligned} a_2^{-1}[x_{(1)}] &= 1 + \sqrt{x_{(1)}} && \text{if } 0 \le x_{(1)} \le 4, \\ &= 3 && \text{if } 4 \le x_{(1)} \le 13. \\ b_2^{-1}[x_{(n)}] &= 3 && \text{if } 0 \le x_{(n)} \le 5, \\ &= 2 - [x_{(n)} - 5]^{1/3}, && \text{if } 5 < x_{(n)} \le 13. \end{aligned}$$

Hence

$$\begin{aligned} T &= 1 + \sqrt{x_{(1)}}, && \text{if } 0 \le x_{(1)} \le 4, \\ & && \phantom{\text{if }} 0 \le x_{(n)} \le 5, \\ &= \min\{1 + \sqrt{x_{(1)}}, 2 - [x_{(n)} - 5]^{1/3}\} && \text{if } 0 \le x_{(1)} \le 4, \\ & && \phantom{\text{if }} 5 < x_{(n)} \le 13, \end{aligned}$$

$= 3$ if $4 \leq x_{(1)} \leq x_{(n)} \leq 5$,

$= 2 - [x_{(n)} - 5]^{1/3}$ if $4 \leq x_{(1)} \leq 13$, $5 < x_{(n)} \leq 13$.

The sampling distribution of T is given by substitution in Eqs. (3-129) and (3-130).

4. Derivation of the General Form of Section 3.10, Assuming the General Form of Equation (2-3), $f(x; \theta) = \exp[u(\theta)\ v(x) + A(x) + B(\theta)]$, When $a \leq x \leq b$, for the Regular Case

Under the assumptions of Theorem 7, as well as the Koopman regularity conditions Eq. (2-3), we prove the conclusion of Theorem 7.

4.1 Necessity Proof

Let a sufficient statistic exist for θ for all $\theta \in \Omega$ for all random samples of any size $n \geq 2$. Then, as remarked in Sec. 2.3, for all $x \in R$ and for all $\theta \in \Omega$, $f(x; \theta)$ is of the form

$$f(x; \theta) = \exp[u(\theta)\ v(x) + A(x) + B(\theta)] \quad [a(\theta) \leq x \leq b(\theta)], \tag{4-1}$$

and the single sufficient statistic must be $T_1 = \sum_{i=1}^{n} v(x_i)$ (or any one-to-one function of T_1). Also, in virtue of the corollary of Theorem 6, $a(\theta)$ and $b(\theta)$ are monotonic in θ in opposite senses, so that condition (1) in the conclusion of Theorem 7 holds good. Then in virtue of Theorem 1 and its extension to monotonic Type II case, T is also a single sufficient statistic for θ for all $\theta \in \Omega$ and for all random samples of any size $n \geq 2$, where T is defined by Eq. (3-146), or (3-147) according as $a(\theta)$ and $b(\theta)$ are monotonic Type I or Type II. Hence both T_1 and T are sufficient for θ for all $\theta \in \Omega$ and for all random samples of any size $n \geq 2$. Then T_1 must be a function of T only (Kendall and Stuart, 1961). But T_1 is a symmetric function of $(x_1, \ldots, x_n)$ and is a sum of separate functions of single variables only, whereas T is a function of $[x_{(1)}, x_{(n)}]$ only.

Hence $T_1 = \sum_{i=1}^{n} v(x_i)$ must be independent of $(x_1, \ldots, x_n)$. That is, $v(x)$ must be a constant k. Then the form (4-1) becomes

$$\begin{aligned} f(x;\ \theta) &= \exp[k\ u(\theta) + A(x) + B(\theta)] \\ &= \exp[A(x)]\ \exp[k\ u(\theta) + B(\theta)] \\ &= g(x)\ h(\theta), \end{aligned}$$

so that condition (2) in the conclusion of Theorem 7 holds good. This completes the necessity proof of the conclusion of Theorem 7.

4.2 Sufficiency Proof

Let the conditions (1) and (2) in the conclusion of Theorem 7 hold good. Let T be the single statistic defined by Eq. (3-146) or (3-147) according as $a(\theta)$ and $b(\theta)$ are monotonic Type I or Type II. Then the range of each x_i for a fixed T is independent of θ, and further the likelihood function of the sample is

$$\begin{aligned} L(x_1, \ldots, x_n;\ \theta) &= [h(\theta)]^n \prod_{i=1}^{n} g(x_i) \\ &= \phi_1(T;\ \theta)\ \phi_2(x_1, \ldots, x_n), \end{aligned} \tag{4-2}$$

where $\phi_1(T;\ \theta) = [h(\theta)]^n$, and $\phi_2(x_1, \ldots, x_n)$ is independent of θ. Hence T satisfied the Factorability Criterion (2-2) in both range and form, and T is therefore a single sufficient statistic for θ. This completes the sufficiency proof.

5. The Nonregular Case with a Single Parameter Admitting a Minimal Set of Jointly Sufficient Statistics

5.1 The Nonregular Case with a Single Parameter Admitting a Minimal Pair of Jointly Sufficient Statistics

In the nonregular case with a single unknown parameter θ, if no single sufficient statistic exists for θ for $n \geq 2$, we can obtain the general form of the family of absolutely continuous distributions admitting a minimal pair of jointly sufficient statistics for θ for all $n \geq 3$. In virtue of the results of Theorems 5 and 7, we make two cases.

CASE I

ASSUMPTIONS. $a(\theta)$ and $b(\theta)$ are continuous, differentiable, and monotonic in θ in opposite senses, and $a(\theta) < b(\theta)$ for all $\theta \varepsilon \Omega$. We further assume the Koopman regularity conditions [form (2-3)] for the regular case of a single sufficient statistic for a single unknown parameter.

Since no single sufficient statistic exists for θ, it follows that under these assumptions, $f(x; \theta)$ cannot be of the form (3-145), for otherwise by Theorem 7 a single sufficient statistic would exist for θ. Let T_1 be the statistic defined by Eq. (3-146) or (3-147) according as $a(\theta)$ and $b(\theta)$ are monotonic Type I or Type II. As shown in Sec. 3, T_1 (or any one-to-one function of T_1) is the only statistic, under the assumptions in question, with the property that the range of each x_i for a fixed T_1 is independent of θ for all $\theta \varepsilon \Omega$. Hence, if a pair of jointly sufficient statistics exists for θ for all $\theta \varepsilon \Omega$ and for all random samples of any size $n \geq 3$, then T_1 (or any one-to-one function of T_1) must be one member of the pair, and without any loss of generality we may take the one-to-one function of T_1 to be T_1 itself. Let T_2 be the other member of the pair of jointly sufficient statistics for θ. Then if T is any other statistic, the conditional distribution of T for a fixed (T_1, T_2) is independent of θ for all $\theta \varepsilon \Omega$ and for all $n \geq 3$. Now the range of T is independent of θ when T_1 alone is fixed. Hence, conditional on T_1 being fixed, the conditional distribution of T for a fixed T_2 is independent of θ in both form and range. That is, conditional on T_1 being fixed, T_2 is a sufficient statistic for θ. Thus, conditional on T_1 being fixed, the range of the distribution of x is independent of θ and the distribution admits a sufficient statistic T_2 for θ. Hence, conditional on T_1 being fixed, we have the regular case (Case I of Sec. 2.3) of the existence of a single sufficient statistic T_2 for the single parameter θ. Let $f(x; \theta|T_1)$ be the p.d.f. of the conditional distribution of x for a fixed T_1. Using Eq. (2-3), the general form of $f(x; \theta|T_1)$ is

$$f(x;\ \theta|T_1) = \exp[u(\theta)\ v(x) + A(x) + B(\theta)]$$

$$[a(\theta) \leq a(T_1) \leq x \leq b(T_1) \leq b(\theta)]. \qquad (5\text{-}1)$$

In Eq. (5-1) $T_2 = \sum_{i=1}^{n} v(x_i)$ is a (conditional) sufficient statistic for θ. Now apart from the extreme observations, T_1 is defined in terms of the terminals a and b only, and the form (5-1) of the p.d.f. $f(x;\ \theta|T_1)$ does not enter into the definition of T_1. Hence the form (5-1) gives also the p.d.f. $f(x;\ \theta)$ of the unconditional distribution of x, i.e.,

$$f(x;\ \theta) = \exp[u(\theta)\ v(x) + A(x) + B(\theta)] \quad [a(\theta) \leq x \leq b(\theta)]. \qquad (5\text{-}2)$$

It may be noted that the other statistics $T_2 = \sum_{i=1}^{n} v(x_i)$ is given by the same expression in both the conditional distribution (5-1) and the unconditional distribution (5-2), since T_2 is defined (apart from the observations) in terms of v(x) only; and the terminals a and b do not enter into the definition of T_2.

In Eq. (5-2) we exclude the particular case when v(x) is constant for all $x \in R$, for in that case Eq. (5-2) will reduce to the form (3-145). Conversely, let $f(x;\ \theta)$ be of the form (5-2), where v(x) is not a constant. Then there exists no single sufficient statistic for θ, since $f(x;\ \theta)$ is not of the form (3-145). Let T_1 be defined by Eq. (3-146) or (3-147) according as $a(\theta)$ and $b(\theta)$ are monotonic Type I or Type II. Let $T_2 = \sum_{i=1}^{n} v(x_i)$. Then the range of each x_i for a fixed $(T_1,\ T_2)$ is independent of θ for all $\theta \in \Omega$ and $n \geq 3$. Also the likelihood function of the sample is

$$L(x_1,\ \ldots,\ x_n;\ \theta) = \exp\left[u(\theta) \sum_{i=1}^{n} v(x_i) + \sum_{i=1}^{n} A(x_i) + n\ B(\theta)\right]$$

$$= \exp[u(\theta)\ T_2 + n\ B(\theta)]\ \exp\left[\sum_{i=1}^{n} A(x_i)\right]$$

$$= \phi_1(T_1,\ T_2;\ \ )\ \phi(x_1,\ \ldots,\ x_n),$$

where $\phi(x_1, \ldots, x_n) = \exp[\sum_{i=1}^{n} A(x_i)]$ is independent of θ. Hence the pair (T_1, T_2) satisfies the Factorability Criterion in both form and range, and is therefore a minimal pair of sufficient statistics for θ. Thus, under the assumptions of Case I, Eq. (5-2), where $v(x)$ is not a constant, gives the general form of the p.d.f. of the family of absolutely continuous distributions admitting a minimal pair of sufficient statistics for θ for $n \geq 3$.

This case belongs to the mixed type (see Sec. 2.10) of nonregular cases. This case is interesting in that one member T_1 of the minimal pair (T_1, T_2) of sufficient statistics depends only on the extreme observations $x_{(1)}$ and $x_{(n)}$, and only the terminals a and b [and not the p.d.f. $f(x; \theta)$] enter into the definition of T_1; whereas the other member T_2 is a symmetric function of all unordered sample members x_i $(i = 1, \ldots, n)$, and only the p.d.f. $f(x; \theta)$ (and not the terminals a and b) enters [through $v(x)$ occurring in the form of $f(x; \theta)$] into the definition of T_2. Also the general form of the p.d.f. $f(x; \theta)$ in Eq. (5-2) is the same as the general form of the p.d.f. $f(x; \theta)$ in Eq. (2-3) for the regular case, whereas the range $[a(\theta), b(\theta)]$ in Eq. (5-2) satisfies the same continuity and monotone restrictions as the range $[a(\theta), b(\theta)]$ in the general form of Sec. 3.10 for the nonregular case. Hence the results obtained in this case are a mixture of the results obtained in the regular and nonregular cases with a single sufficient statistic for a single unknown parameter.

CASE II

ASSUMPTION. For all $\theta \in \Omega$, $a(\theta)$ and $b(\theta)$ are continuous and monotonic in θ, but not in opposite senses; and $a(\theta) < b(\theta)$.

At first it may be remembered that none of $a(\theta)$ and $b(\theta)$ is constant for all $\theta \in \Omega$ in this case. Under these assumptions it follows from Theorem 6 that there exists no single sufficient statistic for θ, and it follows from Theorem 5 that there exists no single statistic T with the property that the range of each x_i for a fixed T is independent of θ for all $\theta \in \Omega$ and $n \geq 2$.

If, however, both $x_{(1)}$ and $x_{(n)}$ are fixed, the range of each x_i is the closed interval $[x_{(1)}, x_{(n)}]$, which is independent of θ for all $\theta \in \Omega$ and $n \geq 3$. The pair $[x_{(1)}, x_{(n)}]$ of statistics [or any pair of one-to-one function of $x_{(1)}$ and $x_{(n)}$] is the only pair having this property. Hence, if a minimal pair of jointly sufficient statistics exists for θ for all $\theta \in \Omega$ and $n \geq 3$ under the assumptions in question, it must be $[x_{(1)}, x_{(n)}]$ or any pair of one-to-one functions of $[x_{(1)}, x_{(n)}]$; and without any loss of generality we may take the pair to be $[x_{(1)}, x_{(n)}]$ itself. By the Form A equivalents of definitions of Sec. 2.2, the pair $[x_{(1)}, x_{(n)}]$ is minimal sufficient for θ for all $\theta \in \Omega$ and $n \geq 3$ if the conditional distribution of x_i for every fixed $[x_{(1)}, x_{(n)}]$ is independent of θ in both range and form for all $\theta \in \Omega$ and $n \geq 3$. The conditional distribution of x_i for a fixed $[x_{(1)}, x_{(n)}]$ $(n \geq 3)$ is obtained by introducing the parameter θ in Eqs. (2-48), (2-50), and (2-51) and considering values of $n > 3$, and is given by

$$P[x_i;\ \theta|x_{(1)}, x_{(n)}] = \frac{n-2}{n} \frac{f(x_i;\ \theta)}{\int_{x_{(1)}}^{x_{(n)}} f(x;\ \theta)\ dx} \qquad [x_{(1)} < x_i < x_{(n)},\ n \geq 3], \tag{5-3}$$

$$P[x_i = x_{(1)};\ \theta|x_{(1)}, x_{(n)}] = \frac{1}{n} \qquad [x_i = x_{(1)} < x_{(n)},\ n \geq 3], \tag{5-4}$$

$$P[x_i = x_{(n)};\ \theta|x_{(1)}, x_{(n)}] = \frac{1}{n} \qquad [x_{(1)} < x_i = x_{(n)},\ n \geq 3]. \tag{5-5}$$

It will be seen from Eqs. (5-3) - (5-5) that the conditional distribution of x_i for every fixed $[x_{(1)}, x_{(n)}]$ $[x_{(1)} \in R,\ x_{(n)} \in R]$ is independent of θ in both range and form for all $\theta \in \Omega$ and $n \geq 3$ if, and only if, $f(x_i;\ \theta)/f(x_j;\ \theta)$ is independent of θ for all $x_i, x_j \in R$ and all $\theta \in \Omega$; that is (in virtue of Sec. 2.5), if, and only if, $f(x;\ \theta)$ is of the form (3-145) for all $x \in R$ and all $\theta \in \Omega$. Hence, under the assumptions of Case II, a necessary and sufficient condition for the existence of a minimal pair of jointly sufficient statistics for θ for all $\theta \in \Omega$ and all $n \geq 3$ is that $f(x;\ \theta)$ is of the form (3-145) for all $x \in R$ and all $\theta \in \Omega$, in which case the pair $[x_{(1)}, x_{(n)}]$ is jointly minimal sufficient for θ.

Since both members of the minimal pair of sufficient statistics are functions of only $[x_{(1)}, x_{(n)}]$, the nonregular case in point belongs to the pure type (Sec. 2.10).

5.2 The Nonregular Case with a Single Parameter Admitting a Minimal Set of $\nu > 2$ Jointly Sufficient Statistics

The general forms in this section can be obtained by straightforward extension of the methods used in Sec. 5.1 (in which $\nu = 2$), and consequently we omit proofs.

CASE I

ASSUMPTIONS. $a(\theta)$ and $b(\theta)$ are continuous, differentiable, and monotonic in θ in opposite senses, and $a(\theta) < b(\theta)$ for all $\theta \in \Omega$. We further assume the Koopman regularity conditions for the general form (2-4) in the regular case, corresponding to a single parameter ($p = 1$) and $\nu - 1$ statistics.

The general form of the p.d.f. $f(x; \theta)$ admitting a minimal set of $\nu > 2$ jointly sufficient statistics for θ for all $\theta \in \Omega$ and $n > \nu$ is

$$f(x;\theta) = \exp\left[\sum_{k=1}^{\nu-1} u_k(\theta)\, v_k(x) + A(x) + B(\theta)\right], \tag{5-6}$$

for all $x \in R$ and $\theta \in \Omega$. In the form (5-6) the $u_k(\theta)$ are supposed to be linearly independent, and so also the v_k. When Eq. (5-6) holds good, the minimal set of ν jointly sufficient statistics for θ is (T_k) $(k = 1, \ldots, \nu)$, where

$$T_k = \sum_{i=1}^{n} v_k(x_i) \qquad (k = 1, \ldots, \nu-1)$$

and T_ν is defined by Eq. (3-146) or (3-147) according as $a(\theta)$ and $b(\theta)$ are monotonic Type I or Type II. It should be noted that in form (5-6), one member of the minimal set of ν sufficient statistics is provided by $[x_{(1)}, x_{(n)}]$, and we assumed the Koopman regularity conditions in relation to the remaining $\nu - 1$ members of the minimal set of sufficient statistics for θ. This case belongs to the mixed type (Sec. 2.10).

CASE II

ASSUMPTIONS. For all $\theta \varepsilon \Omega$, $a(\theta)$ and $b(\theta)$ are continuous and monotonic in θ, but not in opposite senses, and $a(\theta) < b(\theta)$.

The general form of the p.d.f. $f(x; \theta)$ admitting a minimal set of $\nu > 2$ jointly sufficient statistics for θ for all $\theta \varepsilon \Omega$ and $n > \nu$ is

$$f(x;\ \theta) = \exp\left[\sum_{k=1}^{\nu-2} u_k(\theta)\ v_k(x) + A(x) + B(\theta)\right] \tag{5-7}$$

for all $x \varepsilon R$ and $\theta \varepsilon \Omega$. In the form (5-7) the $u_k(\theta)$ $(k = 1, \ldots, \nu-2)$ are supposed to be linearly independent, and so also the $v_k(x)$. When Eq. (5-7) holds good, the minimal set of ν jointly sufficient statistics for θ is (T_k) $(k = 1, \ldots, \nu)$, where

$$T_k = \sum_{i=1}^{n} v_k(x_i) \qquad (k = 1, \ldots, \nu - 2),$$

$$T_{\nu-1} = x_{(1)}, \quad \text{and} \quad T_\nu = x_{(n)}.$$

In the form (5-7), two members of the minimal set of ν sufficient statistics are provided by $x_{(1)}$ and $x_{(n)}$. Consequently the Koopman regularity conditions are assumed in relation to the remaining $\nu - 2$ members of the minimal set of sufficient statistics for θ. Since $\nu > 2$, this case belongs to the mixed type (Sec. 2.10).

6. Derivation of the General Forms of Section 5, Assuming the General Form of Equation (2-4), When $a \leq x \leq b$, for the Regular Case

In the regular case, if no single sufficient statistic exists for a single parameter θ, then under the Koopman regularity conditions the general form of $f(x; \theta)$ admitting a minimal pair of jointly sufficient statistics for θ for $n \geq 3$ is given by putting $p = 1$ and $\nu = 2$ in Eq. (2-4):

$$f(x;\ \theta) = \exp[u_1(\theta)\ v_1(x) + u_2(\theta)\ v_2(x) + A(x) + B(\theta)] \qquad (a \leq x \leq b), \qquad (6\text{-}1)$$

where

$$T_1' = \sum_{i=1}^{n} v_1(x_i) \quad \text{and} \quad T_2' = \sum_{i=1}^{n} v_2(x_i)$$

constitute a minimal pair of jointly sufficient statistics for θ. We now deduce the general forms in Cases I and II of Sec. 5.1, assuming the Koopman regularity conditions in addition to the assumptions made in the respective cases.

6.1 Case I of Section 5.1

Under the additional assumptions of Koopman regularity conditions, the form (6-1) will still be a necessary condition (though not a sufficient condition) for the general form of $f(x;\ \theta)$ in the nonregular case. Now T_1, as defined in Case I of Sec. 5.1, is one member of the minimal pair of jointly sufficient statistics for θ, when such a pair exists. Also, since the form (6-1) is now assumed, (T_1', T_2') is also a minimal pair of jointly sufficient statistics for θ, when such a pair exists. It follows that one member of the pair (T_1', T_2') must be equal to T_1 or a function of T_1 only. To be definite, suppose T_1' is a function of T_1 only. But T_1' is a symmetric function of $(x_1, \ldots, x_n)$ and is a sum of separate functions of single variables only, whereas T_1 depends on $x_{(1)}$ and $x_{(n)}$ only. Hence $v_1(x)$ must be a constant k for all $x \varepsilon R$, in which case Eq. (6-1) reduces to the form

$$f(x;\ \theta) = \exp[u(\theta)\ v(x) + A(x) + B(\theta)] \qquad [a(\theta) \leq x \leq b(\theta)], \qquad (6\text{-}2)$$

where $v(x)$ is not a constant. Thus the form (6-2) is a necessary condition for the existence of a minimal pair of jointly sufficient statistics for θ under the assumptions in question. The sufficiency proof of the form (6-2) is exactly the same as the proof of the converse part in Case I of Sec. 5.1.

6.2 Case II of Section 5.1

In this case also the form (6-1) is a necessary condition for the existence of a minimal pair of jointly sufficient statistics for θ. Also both pairs (T_1', T_2') and $[x_{(1)}, x_{(n)}]$ have a claim for being a minimal pair of jointly sufficient statistics for θ, when such a pair exists. It follows that T_1' and T_2' must be functions of $[x_{(1)}, x_{(n)}]$. But since T_1' and T_2' are symmetric functions of $(x_1, \ldots, x_n)$ and are sums of separate functions of single variables only, we must have $v_1(x) = k_1$ and $v_2(x) = k_2$ for all $x \in R$, where k_1 and k_2 are constants. Hence Eq. (6-1) reduces to the form

$$f(x; \theta) = g(x)\, h(\theta) \qquad [a(\theta) \leq x \leq b(\theta)], \tag{6-3}$$

which is the same form obtained in Case II of Sec. 5.1.

Conversely, if $f(x; \theta)$ is of the form (6-3), then no single sufficient statistic exists for θ under the assumptions made, but the pair $[x_{(1)}, x_{(n)}]$ satisfies the Factorability Criterion in both range and form, so that $[x_{(1)}, x_{(n)}]$ constitutes a minimal pair of jointly sufficient statistics for θ.

REMARK. Though we have assumed the Koopman regularity conditions in the proofs of Case I as given in both Sec. 5.1 and Sec. 6, there is a difference between the assumptions in the two proofs. In the proof of Case I in Sec. 5.1, one member of the pair of minimal sufficient statistics is provided by (i.e., is a function of) $x_{(1)}$ and $x_{(n)}$, and consequently the Koopman regularity conditions are assumed with regard to only the other member of the minimal pair of sufficient statistics, which is not provided by $x_{(1)}$ and $x_{(n)}$, i.e., the Koopman regularity conditions assumed are for the general form (2-3) in the regular case, wherein $p = 1$ and $\nu = 1$. On the other hand, in the proof of Case I in Sec. 6, the Koopman regularity conditions are assumed with regard to both members of the minimal pair of sufficient statistics, i.e., the Koopman regularity conditions assumed are for the general form (2-4) in the regular case with $p = 1$ and $\nu = 2$. Thus the results in Case I are obtained under less stringent conditions in Sec. 5.1 than those in Sec. 6. Moreover, results obtained in Sec. 5.1 for Case II are entirely independent of the Koopman regularity conditions. In like manner, we can obtain alternative derivations of the general forms in Cases I and II of Sec. 5.2, assuming in full the Koopman regularity conditions for the general form (2-4) in the regular case, corresponding to a single parameter ($p = 1$) and ν statistics.

7. The Nonregular Case With Two Parameters

7.1 The Nonregular Case with Two Parameters Admitting a Minimal Pair of Sufficient Statistics

In this section we investigate the general form of the p.d.f. $f(x; \theta_1, \theta_2)$ of the family of absolutely continuous univariate distributions depending on two unknown parameters (θ_1, θ_2) which admit a minimal pair of jointly sufficient statistics for all $n \geq 3$, where the ranges $[a(\theta_1, \theta_2), b(\theta_1, \theta_2)]$ of the distributions depend on (θ_1, θ_2). Let Ω denote the two-dimensional set of admissible values of (θ_1, θ_2). Let Ω_1 and Ω_2 be respectively the one-dimensional sets of admissible values of θ_1 and θ_2. We assume that $a(\theta_1, \theta_2) < b(\theta_1, \theta_2)$ for all $(\theta_1, \theta_2) \in \Omega$. As usual, we shall denote by R the union set of the ranges

$$[a(\theta_1, \theta_2), b(\theta_1, \theta_2)] \quad \text{for } (\theta_1, \theta_2) \in \Omega.$$

We make two cases: in Case I, the range (a, b) depends on both parameters (θ_1, θ_2); in Case II, the range (a, b) depends on only one of the two parameters (θ_1, θ_2), which we can take to be θ_1, without any loss of generality.

It is assumed in Case I that the terminals $a(\theta_1, \theta_2)$ and $b(\theta_1, \theta_2)$ are functionally independent, for otherwise by reparameterization we can have the range (a, b) depending only on a single parameter. In particular, none of the terminals (a, b) is independent of both (θ_1, θ_2). In Case II the p.d.f. must depend on both parameters (θ_1, θ_2), for otherwise the two parameters would not be essential, and a single parameter would then suffice to represent the family of distributions.

CASE I: The range (a, b) depends on both parameters (θ_1, θ_2).

ASSUMPTIONS. We shall not impose any continuity or monotone restrictions on the terminals $a(\theta_1, \theta_2)$ and $b(\theta_1, \theta_2)$. Here the sets Ω, Ω_1, and Ω_2 of admissible values of (θ_1, θ_2), θ_1, and θ_2 respectively can be sets of a general character in their respective dimensions, and need not be intervals.

Since both a and b depend on (θ_1, θ_2), $[x_{(1)}, x_{(n)}]$ is the only pair of statistics [or any pair of one-to-one functions of $x_{(1)}, x_{(n)}$] having the property that, for every $[x_{(1)}, x_{(n)}]$ fixed, the range of each x_i is independent of (θ_1, θ_2) for all $(\theta_1, \theta_2) \in \Omega$ and $n \geq 3$. Hence, if a minimal pair of jointly sufficient statistics exists for (θ_1, θ_2) for all $(\theta_1, \theta_2) \in \Omega$ and $n \geq 3$, it must be $[x_{(1)}, x_{(n)}]$ or any pair of one-to-one functions of $[x_{(1)}, x_{(n)}]$, and without any loss of generality we may take the pair to be $[x_{(1)}, x_{(n)}]$ itself. By the Form A equivalents of definitions of Sec. 2.2, the pair $[x_{(1)}, x_{(n)}]$ is minimal sufficient for (θ_1, θ_2) for all $(\theta_1, \theta_2) \in \Omega$ and $n \geq 3$ if the conditional distribution of x_i for every fixed $[x_{(1)}, x_{(n)}]$ is independent of (θ_1, θ_2) in both range and form for all $(\theta_1, \theta_2) \in \Omega$ and $n \geq 3$. The conditional distribution of x_i for a fixed $[x_{(1)}, x_{(n)}]$ $(n \geq 3)$ is obtained by introducing the parameters θ_1 and θ_2 in Eqs. (2-48), (2-50), and (2-51) and considering values of $n \geq 3$, and is given by

$$P[x_i; \theta_1, \theta_2 | x_{(1)}, x_{(n)}] = \frac{n-2}{n} \frac{f(x_i; \theta_1, \theta_2)}{\int_{x_{(1)}}^{x_{(n)}} f(x; \theta_1, \theta_2)\, dx} \qquad [x_{(1)} < x_i < x_{(n)},\ n \geq 3], \tag{7-1}$$

$$P[x_i = x_{(1)}; \theta_1, \theta_2 | x_{(1)}, x_{(n)}] = \frac{1}{n} \qquad [x_i = x_{(1)} < x_{(n)},\ n \geq 3], \tag{7-2}$$

and

$$p[x_i = x_{(n)}; \theta_1, \theta_2 | x_{(1)}, x_{(n)}] = \frac{1}{n}$$
$$[x_{(1)} < x_i = x_{(n)}, n \geq 3]. \qquad (7\text{-}3)$$

It will be seen from Eqs. (7-1) - (7-3) that the conditional distribution of x_i for every fixed $[x_{(1)}, x_{(n)}]$ $[x_{(1)} \varepsilon R, x_{(n)} \varepsilon R]$ is independent of (θ_1, θ_2) in both range and form for all (θ_1, θ_2) $\varepsilon\ \Omega$ and $n \geq 3$ if, and only if, $f(x_i; \theta_1, \theta_2)/f(x_j; \theta_1, \theta_2)$ is independent of (θ_1, θ_2) for all $x_i, x_j \varepsilon R$ and $(\theta_1, \theta_2) \varepsilon \Omega$; that is (in virtue of the corollary of Sec. 2.5), if, and only if, for all $x \varepsilon R$ and $(\theta_1, \theta_2) \varepsilon \Omega$, $f(x; \theta_1, \theta_2)$ is of the form

$$f(x; \theta_1, \theta_2) = g(x)\ h(\theta_1, \theta_2). \qquad (7\text{-}4)$$

Hence in Case I a necessary and sufficient condition for the existence of a minimal pair of jointly sufficient statistics for (θ_1, θ_2) for all $(\theta_1, \theta_2) \varepsilon \Omega$ and $n \geq 3$ is that $f(x; \theta_1, \theta_2)$ is of the form (7-4) for all $x \varepsilon R$ and $(\theta_1, \theta_2) \varepsilon \Omega$, in which case the pair $[x_{(1)}, x_{(n)}]$ is jointly minimal sufficient for (θ_1, θ_2).

Since both members of the minimal pair of sufficient statistics are functions of only $[x_{(1)}, x_{(n)}]$, the nonregular case in point belongs to the pure type (Sec. 2.10).

CASE II: The range (a, b) depends on the single parameter θ_1.

ASSUMPTIONS. We assume that Ω, Ω_1, and Ω_2 are nondegenerate intervals. $a(\theta_1)$ and $b(\theta_1)$ are continuous, differentiable, and monotonic in θ_1 in opposite senses for all $\theta_1 \varepsilon \Omega_1$. In addition, we assume the Koopman regularity conditions for the general form (2-4) with $p = 2$ and $\nu = 1$.

Let T_1 be the statistic defined by Eq. (3-146) or (3-147) according as $a(\theta)$ and $b(\theta)$ are monotonic Type I or Type II. Then, under the assumptions in question, T_1 is the only single statistic (Sec. 3) with the property that the range of each x_i for a fixed T_1 is independent of θ_1, and hence independent of (θ_1, θ_2) also, for all $(\theta_1, \theta_2) \varepsilon \Omega$ and $n \geq 2$. Hence, if a single sufficient statistic exists for (θ_1, θ_2), it must be T_1 itself (apart from a one-to-one function of T_1). We shall first show that T_1 cannot be a single sufficient statistic for (θ_1, θ_2) for all $(\theta_1, \theta_2) \in \Omega$ and $n \geq 2$. For proceeding exactly as in Sec. 3, we can show that T_1 is a single sufficient statistic for (θ_1, θ_2) for all $(\theta_1, \theta_2) \in \Omega$ and $n \geq 2$ if, and only if, $f(x; \theta_1, \theta_2)$ is of the form

$$f(x; \theta_1, \theta_2) = g(x)\, h(\theta_1, \theta_2) \tag{7-5}$$

for all $x \in R$ and $(\theta_1, \theta_2) \in \Omega$. This is because the assumptions of Sec. 3.1 and the assumptions of the case in question are the same except that the p.d.f. depends on a single parameter θ in Sec. 3, whereas it depends on two parameters (θ_1, θ_2) in the present case. The conditional distribution of x_i for a fixed T_1 for the present case will be simply obtained by writing (θ_1, θ_2) for θ in the conditional distribution of x_i for a fixed T, given by Eqs. (3-108), (3-110), (3-111), and (3-113). Consequently the general form (7-5) is simply obtained by writing (θ_1, θ_2) for θ in the general form (3-114). Since

$$\int_{a(\theta_1)}^{b(\theta_1)} f(x; \theta_1, \theta_2)\, dx = 1 \quad \text{for all } (\theta_1, \theta_2) \in \Omega,$$

Eq. (7-5) implies

$$\int_{a(\theta_1)}^{b(\theta_1)} g(x)\, dx = \frac{1}{h(\theta_1, \theta_2)} \quad \text{for all } (\theta_1, \theta_2) \in \Omega. \tag{7-6}$$

Now the first member of Eq. (7-6) is a function of θ_1 only, whereas the second member is a function of both (θ_1, θ_2). Thus Eq. (7-6) implies that $h(\theta_1, \theta_2)$ is independent of θ_2, i.e., both the p.d.f. and the range of the family of distributions are independent of θ_2, which is contrary to the assumption that the family of distributions depends on two parameters (θ_1, θ_2).

Since T_1 cannot be a single sufficient statistic for (θ_1, θ_2), no single sufficient statistic exists for (θ_1, θ_2) for all

$(\theta_1, \theta_2) \in \Omega$ and $n \geq 2$. We shall now show that if a minimal pair of jointly sufficient statistics exists for (θ_1, θ_2) for all $(\theta_1, \theta_2) \in \Omega$ and $n \geq 3$, then T_1 must be (apart from any one-to-one function of T_1) one member of the pair. For if T_1 is not a member of a minimal pair of jointly sufficient statistics for (θ_1, θ_2) when such a pair exists, such a pair must be $[x_{(1)}, x_{(n)}]$, since $[x_{(1)}, x_{(n)}]$ is the only pair of statistics with the property that the range of each x_i for a fixed $[x_{(1)}, x_{(n)}]$ is independent of (θ_1, θ_2) for all $(\theta_1, \theta_2) \in \Omega$ and $n \geq 3$. Then, proceeding exactly as in Case I of this section, $[x_{(1)}, x_{(n)}]$ is a minimal pair of jointly sufficient statistics for (θ_1, θ_2) if, and only if, $f(x; \theta_1, \theta_2)$ is of the form (7-4), which is the same form (7-5). As we have just shown, the form (7-5) in this case is contrary to the assumption that the family of distributions depends on two parameters (θ_1, θ_2).

We have thus shown that if a minimal pair of jointly sufficient statistics exists for (θ_1, θ_2), then T_1 must be one member of the pair. Suppose now that such a minimal pair exists and let T_2 be the other member of the pair. Then if T is any other statistic, the conditional distribution of T for a fixed (T_1, T_2) is independent of (θ_1, θ_2) for all $(\theta_1, \theta_2) \in \Omega$ and $n \geq 3$. The argument from here onward is exactly the same as that in Case I of Sec. 5.1, except that we have to introduce (θ_1, θ_2) in place of θ in the p.d.f. and the conditional distribution. Hence, replacing θ by (θ_1, θ_2) in the p.d.f. in Eq. (5-2), the general form of $f(x; \theta_1, \theta_2)$ is

$$f(x; \theta_1, \theta_2) = \exp[u(\theta_1, \theta_2)\, v(x) + A(x) + B(\theta_1, \theta_2)] \qquad [a(\theta_1) \leq x \leq b(\theta_1)], \tag{7-7}$$

where $v(x)$ is not a constant. In the form (7-7) the pair of statistics (T_1, T_2), where $T_2 = \sum_{i=1}^{n} v(x_i)$, is minimal sufficient for all $(\theta_1, \theta_2) \in \Omega$ and $n \geq 3$. This nonregular case belongs to the mixed type (Sec. 2.10). The results obtained in this case are a mixture of the results in regular and nonregular cases.

It would appear that we might make another case, a subcase in Case II as the counterpart of Case II just considered, with the assumptions that the range (a, b) depends on the single parameter θ_1 and that $a(\theta_1)$ and $b(\theta_1)$ are continuous and monotonic in θ_1, but not in opposite senses. But in such a case there exists no single statistic with the property that when it is fixed, the range of each x_i is independent of θ_1 and hence independent of (θ_1, θ_2) also. Hence there exists no single sufficient statistic for (θ_1, θ_2). Then if a minimal pair of jointly sufficient statistics exists for (θ_1, θ_2), it must be $[x_{(1)}, x_{(n)}]$, and this would imply (as shown in Case II of this section) that $f(x; \theta_1, \theta_2)$ is of the form (7-5), which is not permissible in this case for the reasons already mentioned.

7.1.1 Derivation of the general form of Equation (7-4), $f(x; \theta_1, \theta_2) = g(x)\, h(\theta_1, \theta_2)$, of Case I of Section 7.1, assuming the general form of Equation (2-4),

$$f(x; \theta_1, \ldots, \theta_p) = \exp\left[\sum_{k=1}^{\nu} u_k(\theta_1, \ldots, \theta_p)\, v_k(x) + A(x) + B(\theta_1, \ldots, \theta_p)\right],$$

when $a \leq x \leq b$, for the regular case with $p = 2$ and $\nu = 2$

If we assume the Koopman regularity conditions in Case I of Sec. 7.1, then a necessary condition (though not a sufficient condition) for the existence of a minimal pair of jointly sufficient statistics for (θ_1, θ_2) is that $f(x; \theta_1, \theta_2)$ should be of the form (2-4) with $p = 2$ and $\nu = 2$. Hence

$$f(x; \theta_1, \theta_2) = \exp[u_1(\theta_1, \theta_2)\, v_1(x) + u_2(\theta_1, \theta_2)\, v_2(x) + A(x) + B(\theta_1, \theta_2)]$$
$$[a(\theta_1, \theta_2) \leq x \leq b(\theta_1, \theta_2)]. \quad (7\text{-}8)$$

Then if $T_1' = \sum_{i=1}^{n} v_1(x_i)$ and $T_2' = \sum_{i=1}^{n} v_2(x_i)$, both the pairs $[x_{(1)}, x_{(n)}]$ and (T_1', T_2') have a claim for being a minimal pair of jointly sufficient statistics for (θ_1, θ_2), if such a pair exists. This would imply that $v_1(x) = k_1$ and $v_2(x) = k_2$, where k_1 and k_2 are constants, and the form (7-8) reduces to the form (7-4). The sufficiency proof of the form (7-4) is completed when it is seen that when $f(x; \theta_1, \theta_2)$ is of the form (7-4), the pair $[x_{(1)}, x_{(n)}]$ satisfies the Factorability Criterion in both range and form.

Similarly the form (7-8) may also be used to obtain alternative proofs of some results in Case II of Sec. 7.1. But these alternative proofs involve more stringent assumptions.

7.1.2 Illustrative examples

EXAMPLE 1. Consider the two-parameter rectangular distribution with p.d.f.

$$f(x; \mu, \sigma) = \frac{1}{\sigma} \quad [\mu - \tfrac{1}{2}\sigma \le x \le \mu + \tfrac{1}{2}\sigma].$$

Here the range (a, b) depends on both parameters μ and σ, and by Case I of Sec. 7.1, $[x_{(1)}, x_{(n)}]$ is a minimal pair of jointly sufficient statistics for (μ, σ). The natural ranges of admissible values of μ and σ are the intervals $(-\infty, \infty)$ and $(0, \infty)$ respectively. However, $[x_{(1)}, x_{(n)}]$ will still be a minimal pair of jointly sufficient statistics for (μ, σ) even if the ranges of μ and σ are respectively any subsets of $(-\infty, \infty)$ and $(0, \infty)$, since the ranges need not be intervals in Case I. This example belongs to the pure-type nonregular case.

EXAMPLE 2. Consider the two-parameter exponential distribution with p.d.f.

$$f(x; \mu, \sigma) = \frac{1}{\sigma}\exp\left(-\frac{x - \mu}{\sigma}\right) \qquad (\mu \le x < \infty).$$

This is an illustration of Case II of Sec. 7.1, since the range (μ, ∞) depends only on the single parameter μ. For Ω_1 the range of μ is $(-\infty, \infty)$; for Ω_2 the range of σ is $(0, \infty)$. Here $a(\mu) = \mu$ is m.i. and $b(\mu) = \infty$ is independent of μ. Then Eq. (3-151) gives

$$T_1 = a_2^{-1}[x_{(1)}] = x_{(1)}.$$

$f(x; \mu, \sigma)$ is of the form (7-7) with $v(x) = x$:

$$\sum_{i=1}^{n} v(x_i) = n\bar{x},$$

where $\bar{x}$ is the sample mean. We may then take

$$T_2 = \frac{1}{n} \sum_{i=1}^{n} v(x_i) = \bar{x}.$$

Hence the pair $[x_{(1)}, \bar{x}]$ is minimal sufficient for (μ, σ). This example belongs to the mixed-type nonregular case. Similarly, it can be shown that for the other two-parameter exponential distribution

$$f(x;\ \mu, \sigma) = \frac{1}{\sigma} \exp\left(- \frac{\mu - x}{\sigma}\right) \qquad (-\infty < x \leq \mu),$$

the pair $[x_{(n)}, \bar{x}]$ of statistics is minimal sufficient for (μ, σ).

7.2 The Nonregular Case with Two Parameters Admitting a Minimal Set of $\nu > 2$ Jointly Sufficient Statistics

The general forms in this section can be obtained by straightforward extension of the methods used in Sec. 7.1 (in which $\nu = 2$) as well as the methods used in Sec. 5, and hence we shall omit proofs.

CASE I: The range (a, b) depends on both parameters (θ_1, θ_2).

ASSUMPTIONS. Let the admissible values of (θ_1, θ_2) belong to some two-dimensional nondegenerate interval Ω. Let the set R be the union of the ranges $[a(\theta_1, \theta_2), b(\theta_1, \theta_2)]$ for all $(\theta_1, \theta_2) \in \Omega$. We assume the Koopman regularity conditions for the general form (2-4) in the regular case, corresponding to $p = 2$ parameters and $\nu - 2$ statistics.

The general form of the p.d.f. $f(x;\ \theta_1, \theta_2)$ admitting a minimal set of $\nu > 2$ jointly sufficient statistics for (θ_1, θ_2) for all $(\theta_1, \theta_2) \in \Omega$ and $n > \nu$ is

$$f(x;\ \theta_1, \theta_2) = \exp\left[\sum_{k=1}^{\nu-2} u_k(\theta_1, \theta_2)\ v_k(x) + A(x) + B(\theta_1, \theta_2)\right] \tag{7-9}$$

for all $x \in R$ and $(\theta_1, \theta_2) \in \Omega$. In the form (7-9) the $u_k(\theta_1, \theta_2)$

$(k = 1, \ldots, \nu - 2)$ are supposed to be linearly independent, and so also the $v_k(x)$. When Eq. (7-9) holds good, the minimal set of ν jointly sufficient statistics for (θ_1, θ_2) is (T_k) $(k = 1, \ldots, \nu)$, where

$$T_k = \sum_{i=1}^{n} v_k(x_i) \qquad (k = 1, \ldots, \nu - 2),$$

$$T_{\nu-1} = x_{(1)}, \quad \text{and} \quad T_\nu = x_{(n)}.$$

In the form (7-9) two members of the minimal set of ν sufficient statistics are provided by $x_{(1)}$ and $x_{(n)}$, and consequently the Koopman regularity conditions are assumed in relation to the remaining $\nu - 2$ members of the minimal set of sufficient statistics for (θ_1, θ_2).

CASE II: The range (a, b) depends on the single parameter θ_1.

We divide this case into two subcases.

SUBCASE 1

ASSUMPTIONS. Let the admissible values of (θ_1, θ_2) belong to some two-dimensional nondegenerate interval Ω. Let Ω_1 and Ω_2 be respectively one-dimensional nondegenerate intervals of admissible values of θ_1 and θ_2. R has the same meaning as in Case I of this section. $a(\theta_1)$ and $b(\theta_1)$ are continuous, differentiable, and monotonic in θ_1 in opposite senses for all $\theta_1 \in \Omega_1$. In addition we assume the Koopman regularity conditions for the general form (2-4) in the regular case, corresponding to $p = 2$ parameters and $\nu - 1$ statistics.

The general form of the p.d.f. $f(x; \theta_1, \theta_2)$ admitting a minimal set of $\nu > 2$ jointly sufficient statistics for (θ_1, θ_2) for all $(\theta_1, \theta_2) \in \Omega$ and $n > \nu$ is

$$f(x; \theta_1, \theta_2) = \exp\left[\sum_{k=1}^{\nu-1} u_k(\theta_1, \theta_2)\, v_k(x) + A(x) + B(\theta_1, \theta_2)\right], \tag{7-10}$$

for all $x \in R$ and $(\theta_1, \theta_2) \in \Omega$. In the form (7-10) the $u_k(\theta_1, \theta_2)$ $(k = 1, \ldots, \nu - 1)$ are linearly independent, and so also the $v_k(x)$.

When Eq. (7-10) holds good, the minimal set of ν jointly sufficient statistics for (θ_1, θ_2) is (T_k) $(k = 1, \ldots, \nu)$, where

$$T_k = \sum_{i=1}^{n} v_k(x_i) \qquad (k = 1, \ldots, \nu - 1),$$

and T_ν is defined by Eq. (3-146) or (3-147) according as $a(\theta_1)$ and $b(\theta_1)$ are monotonic Type I or monotonic Type II. In the form (7-10) one member of the minimal set of ν jointly sufficient statistics is provided by $[x_{(1)}, x_{(n)}]$ and hence the Koopman regularity conditions are assumed in relation to the remaining $\nu - 1$ members of the minimal set of jointly sufficient statistics for (θ_1, θ_2).

SUBCASE 2

ASSUMPTIONS. Let Ω, Ω_1, Ω_2, and R have the same meanings as in Subcase 1. For all $\theta_1 \in \Omega_1$, $a(\theta_1)$ and $b(\theta_1)$ are continuous and monotonic in θ_1, but not in opposite senses. In addition, we assume the Koopman regularity conditions for the general form (2-4) for the regular case, corresponding to $p = 2$ parameters and $\nu - 2$ statistics.

The possibility of such a subcase was discussed (see the concluding remarks in Case II of Sec. 7.1) for the case $\nu = 2$ and was found to be inadmissible in that case. However, such a subcase can arise if $\nu > 2$. Using the argument in the conclusing remarks in Case II of Sec. 7.1, it can be shown that the general form of $f(x; \theta_1, \theta_2)$ for this subcase is exactly the same as the general form (7-9) in Case I of Sec. 7.2, and the minimal set of jointly sufficient statistics in Case I is also the minimal set of jointly sufficient statistics in this subcase. Case I and this subcase differ only in the form of the range (a, b), which depends on both (θ_1, θ_2) in Case I, and only on θ_1 in this subcase.

It may be noted that all cases considered in Sec. 7.2 belong to the mixed type.

Finally it may be remarked that the forms in Sec. 7.2 may also be derived under more stringent conditions by assuming in full the Koopman regularity conditions for the general form (2-4) in the regular case, corresponding to $p = 2$ parameters and ν statistics.

8. The Nonregular Case Involving More Than Two Parameters

In the nonregular case, if a family of distributions depends on $p > 2$ parameters $(\theta_1, \ldots, \theta_p)$, then by reparameterization we can have the range (a, b) of distributions depending on at most two parameters, which we take as (θ_1, θ_2). Let the admissible values of $(\theta_1, \ldots, \theta_p)$ belong to some p-dimensional nondegenerate interval Ω. Let Ω_i be the one-dimensional nondegenerate interval of admissible values of θ_i $(i = 1, \ldots, p)$. Let the set R be the union of the ranges $[a(\theta_1, \theta_2), b(\theta_1, \theta_2)]$ for all $\theta_1 \in \Omega_1$ and $\theta_2 \in \Omega_2$. The derivation of the general forms in this section is exactly on similar lines as the derivation of the general forms in Sec. 7.2. Consequently we shall omit proofs. We consider the following cases:

CASE I: The range (a, b) depends on both parameters (θ_1, θ_2).

ASSUMPTIONS. We assume the Koopman regularity conditions for the general form (2-4) in the regular case, corresponding to p parameters and $\nu - 2$ statistics. The general form of the p.d.f. $f(x; \theta_1, \ldots, \theta_p)$ admitting a minimal set of ν jointly sufficient statistics for $(\theta_1, \ldots, \theta_p)$ for all $(\theta_1, \ldots, \theta_p) \in \Omega$ and $n > \nu$ is

$$f(x; \theta_1, \ldots, \theta_p) = \exp\left[\sum_{k=1}^{\nu-2} u_k(\theta_1, \ldots, \theta_p)\, v_k(x) + A(x) + B(\theta_1, \ldots, \theta_p)\right] \tag{8-1}$$

for all $x \in R$ and $(\theta_1, \ldots, \theta_p) \in \Omega$. In Eq. (8-1) the u_k $(k = 1, \ldots, \nu - 2)$ are supposed to be linearly independent, and so also the v_k. When Eq. (8-1) holds good, the minimal set of ν jointly sufficient statistics is (T_k) $(k = 1, \ldots, \nu)$, where

$$T_k = \sum_{i=1}^{n} v_k(x_i) \quad (k = 1, \ldots, \nu - 2),$$
$$T_{\nu-1} = x_{(1)}, \quad \text{and} \quad T_\nu = x_{(n)}. \tag{8-2}$$

CASE II: The range (a, b) depends on the single parameter θ_1.

We divide this case into two subcases.

SUBCASE 1

ASSUMPTIONS. $a(\theta_1)$ and $b(\theta_1)$ are continuous, differentiable, and monotonic in θ_1 in opposite senses for all $\theta_1 \in \Omega_1$. In addition we assume the Koopman regularity conditions for the general form (2-4) in the regular case, corresponding to p parameters and $\nu - 1$ statistics. The general forms of the p.d.f. $f(x; \theta_1, \ldots, \theta_p)$ admitting a minimal set of ν jointly sufficient statistics for $(\theta_1, \ldots, \theta_p)$ for all $(\theta_1, \ldots, \theta_p) \in \Omega$ and $n > \nu$ is

$$f(x; \theta_1, \ldots, \theta_p) = \exp\left[\sum_{k=1}^{\nu-1} u_k(\theta_1, \ldots, \theta_p)\, v_k(x) + A(x) + B(\theta_1, \ldots, \theta_p)\right], \qquad (8\text{-}3)$$

for all $x \in R$ and $(\theta_1, \ldots, \theta_p) \in \Omega$. In Eq. (8-3) the u_k $(k = 1, \ldots, \nu - 1)$ are supposed to be linearly independent, and so also the v_k. When Eq. (8-3) holds good, the minimal set of ν jointly sufficient statistics is (T_k) $(k = 1, \ldots, \nu)$, where

$$T_k = \sum_{i=1}^{n} v_k(x_i) \quad (k = 1, \ldots, \nu - 1)$$

and T_ν is defined by Eq. (3-146) or (3-147) according as $a(\theta_1)$ and $b(\theta_1)$ are monotonic Type I or monotonic Type II.

SUBCASE 2

ASSUMPTIONS. For all $\theta_1 \in \Omega_1$, $a(\theta_1)$ and $b(\theta_1)$ are continuous and monotonic in θ_1, but not in opposite senses. In addition we assume the Koopman regularity conditions for the general form (2-4) for the regular case, corresponding to p parameters and $\nu - 2$ statistics. The general form $f(x; \theta_1, \ldots, \theta_p)$ for this subcase is exactly the same as the general form (8-1) in Case I of this section, and the set of statistics (8-2) is also the minimal set of jointly sufficient statistics in this subcase. Case I and this subcase differ only in the form of the range (a, b), which depends on both (θ_1, θ_2) in Case I, and only on θ_1 in this subcase. The forms in Case I and Case II of this section may also be deduced under more stringent conditions by assumption of the Koopman regularity conditions for the general form (2-4) in the regular case, corresponding to p parameters and ν statistics.

Finally it may be remarked here that all cases considered in this section belong to the mixed type.

ACKNOWLEDGMENTS

I am exceedingly grateful to the referee for making several useful suggestions toward improvement of this discussion. I wish to express my sincerest thanks to the authorities of the Iowa State University and to Professor T. A. Bancroft, Director of the Statistical Laboratory, for help and facilities for research work.

This work was supported by the National Science Foundation (N.S.F. GP-1155) and was written when the author was Fulbright Visiting Professor at Iowa State University.

REFERENCES

Cramer, H. (1946): Mathematical Methods of Statistics, Princeton Univ. Press, 57.

Davis, R. C. (1951): On minimum variance in nonregular estimation, Ann. Math. Stat. 22, 43.

Darmois, G. (1935): Sur les lois de probabilité à estimation exhaustive, C. R. Acad. Sci. Paris 200, 1265.

Halmos, P. R., and Savage, L. J. (1949): Application of the Radon-Nikodym theorem to the theory of sufficient statistics, Ann. Math. Stat. 20, 225.

Huzurbazar, V. S. (1955): Confidence intervals for the parameter of a distribution admitting a sufficient statistic when the range depends on the parameter, J. R. Stat. Soc. B 17, 86.

Kendall, M. G., and Stuart, A. (1961): The Advanced Theory of Statistics, Vol. 2, London, Charles Griffin and Co.

Koopman, B. O. (1936): On distributions admitting a sufficient statistic, Trans. Amer. Math. Soc. 39, 399.

Neyman, J. (1935): Sur un teorema concernente le cosidette statistiche sufficienti, Giorn. Ist. Ital. Att. 6, 320.

Pitman, E. J. G. (1936): Sufficient statistics and intrinsic accuracy, Proc. Camb. Phil. Soc. 32, 567.

PART 3

LOCATION AND SCALE PARAMETERS, AND SUFFICIENT STATISTICS

1. Introduction

1.1 General Introduction to Results Obtained

The main object of this discussion is to investigate properties of univariate distributions depending on location or scale parameters, or on both parameters of location and scale, and then show under fairly general conditions that all the well-known location-parameter, or scale-parameter, or location-scale-parameter families of univariate distributions are characterized by combinations of the properties of location, scale, and sufficiency: (1) location and sufficiency (2) scale and sufficiency, and (3) location, scale, and sufficiency.

Section 2 deals with definitions, notation, and preliminary results. In Sec. 2.1 a precise definition of the range of a probability distribution is given. This enables us to define formally the regular and nonregular cases of estimation of unknown parameters. Sections 2.4 and 2.5 give statements of the general forms of univariate distributions admitting sufficient statistics for parameters in regular and nonregular cases. Rigorous proofs and discussion of the general forms in nonregular cases have been given (Huzurbazar, 1964) in another paper. Section 2.6 gives statements of general solutions of some known generalized Cauchy functional equations which occur frequently in this paper. Section 2.7 gives the general solution of a new functional equation which plays an important role in this paper.

Section 3 deals with distributions depending on location parameter only. A formal definition of a location parameter is given, and it is shown that the natural range of a location parameter is the interval $R_1 = (-\infty, \infty)$. In Sec. 3.4 it is shown that the range of a location-parameter family of distributions of the regular type is R_1, which is also the range of the location parameter. For a location-parameter family of distributions of the nonregular type, the terminals of the ranges differ from the location parameter by additive constants only; and it is also shown that the ranges belong to one of the three types, given by

Eqs. (3-6) - (3-8). In Secs. 3.5 - 3.7 it is shown that in a variate transformation or a parameter transformation or both, a location-parameter family of distributions will continue to be a location-parameter family of distributions if, and only if, each transformation is linear with change of origin but without change of scale.

Section 4 depends on distributions depending on scale parameter only. Starting from a formal definition of a scale parameter, it is shown that the natural range of a scale parameter is the open interval $(0, \infty)$. It is shown in Sec. 4.4 that the range of a scale-parameter family of distributions of the regular type is either $(-\infty, \infty)$ or $(-\infty, 0)$ or $(0, \infty)$. For a scale-parameter family of distributions of the nonregular type, the terminals of the ranges are multiples of the scale parameter; and it is shown that the ranges belong to one of the five types given by Eqs. (4-5) - (4-9). In Secs. 4.5 - 4.7 it is shown that in a variate transformation or a parameter transformation or both, a scale-parameter family of distributions will continue to be a scale-parameter family of distributions if, and only if, each transformation is linear with change of scale but without change or origin.

Section 5 is concerned with distributions depending on both location and scale parameters. It is shown that the range of a location-scale-parameter family of distributions of the regular type is R_1. For a location-scale-parameter family of distributions of the nonregular type, the ranges belong to one of the seven different types given by Eqs. (5-6) - (5-12). The interesting point to be noted here is that either the ranges depend on both parameters (θ, σ) or they depend on the location parameter θ only. Further, if the ranges depend on the single parameter θ, the ranges must be either $(-\infty, \theta)$ or (θ, ∞). In Secs. 5.4 - 5.6 it is shown that in a variate transformation or a parameter transformation or both, a location-scale-parameter family of distributions will continue to be a location-scale-parameter family of distributions if, and only if, the transformations are linear and belong to the type (5-47).

It is shown in Sec. 6 that the family of normal distributions (with known scale parameter), and the family of generalized iterated exponential (also called "double-exponential" or "extreme-value") distributions (with known scale and numerical parameters) are the only location-parameter families of distributions of the regular type admitting a single sufficient statistic for the location parameter for all random samples of any size $n \geq 2$. The main assumption under which these results are obtained is the existence of the third derivative of the p.d.f. (probability density function) of the location-parameter family of distributions.

Section 7 gives the scale-parameter families of distributions of the regular type admitting a single sufficient statistic for the scale parameter for all random samples of any size $n \geq 2$. The main assumption is the existence of the third derivative of the p.d.f. of the scale-parameter families of distributions. The scale-parameter families of distributions obtained include also the family of normal distributions (with known location) as a particular case.

Section 8 is devoted to the characterization of the two-parameter univariate normal distribution. Under the main assumption of the existence of the third derivative of the p.d.f., it is shown that the family of normal distributions is the only location-scale-parameter family of distributions of the regular type admitting a minimal pair of jointly sufficient statistics for the parameters of location and scale for all random samples of any size $n \geq 3$.

Under the assumption of the existence of the first derivative of the p.d.f., it is shown in Sec. 9 that the only location-parameter families of distributions of the nonregular type admitting a single sufficient statistic for the location parameter θ for all random samples of any size $n \geq 2$ are the two exponential families of distributions with ranges $(-\infty, \theta)$ and (θ, ∞) and known scale parameter σ. This nonregular case belongs to the pure type (Sec. 2.3).

Section 10 is concerned with derivation of the scale-parameter families of distributions of the nonregular type admitting a single sufficient statistic for the scale parameter for all random samples of any size $n \geq 2$. The families derived include also the scale-parameter families of rectangular distributions as special cases. This nonregular case also belongs to the pure type.

Section 11 is concerned with the characterization of the two-parameter rectangular and exponential families of distributions. Under the assumption of the existence of the first derivative of the p.d.f., it is shown that if the ranges of a location-scale-parameter family of distributions of the nonregular type depend on both location and scale parameters, then the rectangular family of distributions is the only location-scale parameter family of distributions admitting a minimal pair of jointly sufficient statistics for the parameters of location and scale for all random samples of any size $n \geq 3$. This nonregular subcase also belongs to the pure type. On the other hand, if the ranges of a location-scale-parameter family of distributions of the nonregular type depend on a single parameter only, then under the main assumption of the existence of the second derivative of the p.d.f., it is shown that the exponential family of distributions is the only location-scale-parameter family of distributions admitting a minimal pair of jointly sufficient statistics for the parameters of location and scale for all random samples of any size $n \geq 3$. This nonregular subcase belongs to the mixed type.

Finally it may be remarked that in characterization of families of distributions, existence of only the first derivative of the p.d.f. is assumed in pure-type nonregular cases, existence of the second derivative of the p.d.f. is assumed in the mixed-type nonregular case, and existence of the third derivative of the p.d.f. is assumed in regular cases of estimation.

2. Definitions, Notation, and Preliminary Results

2.1 Definition of the Range of a Probability Distribution

Let $F(x)$ be the d.f. (distribution function) of a one-dimensional probability distribution. Although $F(x)$ is defined over the interval $R_1 = (-\infty, \infty)$, and although R_1 is often spoken of as the range of the probability distribution, we shall use the term "range of a probability distribution" in the sense of the smallest closed interval (a, b) having the property

$$P(a, b) = 1, \tag{2-1}$$

where the symbol $P(a, b)$ denotes the probability measure of the closed interval (a, b). In special cases either a may be $-\infty$ or b may be $+\infty$ or both may hold good. It can be easily shown that the range (a, b) as defined by Eq. (2-1) is unique. Further, a is given by

$$\begin{aligned} &F(x) = 0 \quad \text{for all } x < a \\ \text{and} \quad &F(x) > 0 \quad \text{for all } x > a, \end{aligned} \tag{2-2}$$

and b is given by

$$\begin{aligned} &F(x) < 1 \quad \text{for all } x < b \\ \text{and} \quad &F(x) = 1 \quad \text{for all } x > b. \end{aligned} \tag{2-3}$$

If the distribution possesses p.d.f. $f(x)$, then a is given by

$$\begin{aligned} &f(x) = 0 \quad \text{for all } x < a, \\ \text{and} \quad &f(x) > 0 \quad \text{for at least one } x \text{ belonging to the} \\ &\qquad\qquad \text{neighborhood } (a, a + \varepsilon) \text{ of } a \text{ for every } \varepsilon > 0. \end{aligned} \tag{2-4}$$

Similarly b is given by

$$\begin{aligned} &f(x) = 0 \quad \text{for all } x < b, \\ \text{and} \quad &f(x) > 0 \quad \text{for at least one } x \text{ belonging to the} \\ &\qquad\qquad \text{neighborhood } (b - \varepsilon, b) \text{ of } b \text{ for every } \varepsilon > 0. \end{aligned} \tag{2-5}$$

2.2 Definitions of the Regular and Nonregular Cases of Estimation of Unknown Parameters

Consider a family of distributions depending on some unknown parameters. If the ranges (as defined in Sec. 2.1) of the family of distributions are independent of the parameters (i.e., the range is the same for every member of the family of distributions), we shall call it a regular case of estimation. If the ranges depend on the parameters, we shall call it a nonregular case of estimation.

2.3 Definitions of Pure-type and Mixed-type Cases in Search for Sufficient Statistics for Parameters in Nonregular Cases

A nonregular case admitting jointly sufficient statistics for all parameters will be said to be of the "pure type" if every member of the minimal set of sufficient statistics is a function of only the smallest and the greatest members in the sample.

A nonregular case admitting jointly sufficient statistics for all parameters will be said to be of the "mixed type" if it is not of the pure type.

2.4 Statements of the General Forms of Univariate Distributions Admitting Sufficient Statistics for Parameters in Regular Cases

For the purposes of this discussion we shall mainly require the general forms in two cases:

CASE I: The regular case with a single sufficient statistic for a single unknown parameter

Let $f(x; \theta)$ be the p.d.f. of a family of distributions depending on a single unknown parameter θ whose admissible values belong to some nondegenerate interval Ω. Let the range (a, b) of the distribution of x be independent of θ. Under certain conditions the general form of $(x; \theta)$ admitting a single sufficient statistic for θ for all $\theta \in \Omega$ and for all random samples of any size $n \geq 2$ is

$$f(x; \theta) = \exp[u(\theta)\ v(x) + A(x) + B(\theta)], \quad (a \leq x \leq b) \qquad (2\text{-}6)$$

where u and B are functions of θ only, and v and A are functions of x only. The form (2-6) was given by Koopman (1936), Pitman (1936), and Darmois (1935). For a statement of the set of conditions under which the form (2-6) is obtained, see Koopman (1936). We shall use the phrase "The Koopman regularity conditions" for conditions under which the form (2-6) is derived. The Koopman regularity conditions assume mainly the existence of the partial derivative $\partial^2 f/(\partial\theta\ \partial x)$ of the second order.

It may be remarked that if n = 1, the term "general form" becomes meaningless, since in that case every distribution possesses a single sufficient statistic for a single unknown parameter.

CASE II: The regular case with two unknown parameters admitting a minimal pair of jointly sufficient statistics

In the regular case, if a family of distributions with p.d.f. $f(x; \theta_1, \theta_2)$ depends on two unknown parameters (θ_1, θ_2) where the admissible values of (θ_1, θ_2) belong to some two-dimensional non-degenerate interval Ω, then under the Koopman regularity conditions the general form of $f(x; \theta_1, \theta_2)$ admitting a minimal pair of jointly sufficient statistics for (θ_1, θ_2) for all $(\theta_1, \theta_2) \in \Omega$ and for all random samples of any size $n \geq 3$ is

$$f(x; \theta_1, \theta_2) = \exp\Big[u_1(\theta_1, \theta_2)\ v_1(x) + u_2(\theta_1, \theta_2)\ v_2(x) + A(x) + B(\theta_1, \theta_2)\Big] \qquad (a \leq x \leq b), \quad (2\text{-}7)$$

where u_1 and u_2 are linearly independent functions of (θ_1, θ_2) only, B is a function of (θ_1, θ_2) only, v_1 and v_2 are linearly independent functions of x only, and A is a function of x only. The Koopman regularity conditions in this case assume mainly the existence of the second-order partial derivatives $\partial^2 f/(\partial\theta_1\ \partial x)$ and $\partial^2 f/(\partial\theta_2\ \partial x)$.

It may be remarked that if $n \leq 2$, the term "general form" becomes meaningless, since in that case every distribution admits a pair of jointly sufficient statistics for a pair of unknown parameters.

The regular case with several unknown parameters. Let $f(x; \theta_1, \ldots, \theta_p)$ be the p.d.f. of a family of distributions depending on p unknown parameters $(\theta_1, \ldots, \theta_p)$. Under the Koopman regularity conditions, the general form of $f(x;\ \theta_1, \ldots, \theta_p)$ admitting a minimal set of ν jointly sufficient statistics for $(\theta_1, \ldots, \theta_p)$ for all random samples of any size $n > \nu$ is

$$f(x;\ \theta_1, \ldots, \theta_p) = \exp\left[\sum_{k=1}^{\nu} u_k(\theta_1, \ldots, \theta_p)\ v_k(x) + A(x) + B(\theta_1, \ldots, \theta_p)\right], \qquad (2\text{-}8)$$

where the u_k $(k = 1, \ldots, \nu)$ are linearly independent functions of $(\theta_1, \ldots, \theta_p)$, and the v_k $(k = 1, \ldots, \nu)$ are linearly independent functions of x. ν may be greater than, equal to, or smaller than p.

p.

2.5 Statements of the General Forms of Univariate Distributions Admitting Sufficient Statistics for Parameters in Nonregular Cases

In the nonregular case Pitman (1936) gives the general forms of univariate distributions with a single unknown parameter admitting a single or a minimal pair of sufficient statistics. There are inaccuracies, incompleteness, and conjecture in Pitman's derivation of the general forms in question; and moreover, the assumptions have not been clearly stated by him. For rigorous proofs and discussion of the nonregular cases with a single or several unknown parameters admitting sufficient statistics, see Huzurbazar (1964). For the purposes of this discussion, however, we shall require the general forms in two nonregular cases only:

CASE I: The nonregular case with a single unknown parameter admitting a single sufficient statistic

Assumptions. Let $f(x;\ \theta)$ be the p.d.f. of the family of distributions depending on a single unknown parameter θ and with ranges $[a(\theta), b(\theta)]$, where the range of admissible values of θ is some nondegenerate interval Ω which may be closed, open, half-open, finite, or infinite. Let the set R be the union of the ranges $[a(\theta), b(\theta)]$ for all $\theta \in \Omega$. For all $\theta \in \Omega$, $a(\theta)$ and $b(\theta)$ are continuous, differentiable, monotonic functions of θ, and $a(\theta) < b(\theta)$.

Under these assumptions, for the existence of a single sufficient statistic for θ for all $\theta \in \Omega$ for all random samples of any size $n \geq 2$, it is both necessary and sufficient that (1) $a(\theta)$ and $b(\theta)$ are monotonic in θ in opposite senses, and that (2) for all $x \in R$ and all $\theta \in \Omega$, $f(x; \theta)$ is of the form

$$f(x; \theta) = g(x)\, h(\theta), \tag{2-9}$$

where $g(x)$ is a function of x defined in R, and $h(\theta)$ is a function of θ defined in Ω.

In the necessary and sufficient condition (1), the phrase "monotonic in θ in opposite senses" means that either $a(\theta)$ is m.i. (monotone increasing) and $b(\theta)$ is m.d. (monotone decreasing), or else $a(\theta)$ is m.d. and $b(\theta)$ is m.i.

When conditions (1) and (2) are satisfied, the single sufficient statistic T for θ is a function of $[x_{(1)}, x_{(n)}]$, where $x_{(1)}$ and $x_{(n)}$ are respectively the smallest and the greatest members in the sample. For a general formula for T covering different cases, see Huzurbazar (1964). This nonregular case belongs to the pure type, since the single sufficient statistic T is a function of $[x_{(1)}, x_{(n)}]$.

CASE II: The nonregular case with two unknown parameters (θ_1, θ_2) admitting a minimal pair of jointly sufficient statistics

This case is divided into two sub-cases:

SUBCASE 1: The ranges (a, b) depend on both parameters (θ_1, θ_2).

Let $f(x; \theta_1, \theta_2)$ be the p.d.f. of the family of distributions. Let Ω denote the two-dimensional set of admissible values of (θ_1, θ_2) and let the set R be the union of the ranges $[a(\theta_1, \theta_2), b(\theta_1, \theta_2)]$ for all $(\theta_1, \theta_2) \in \Omega$. No monotone or continuity restrictions are imposed on the terminals $a(\theta_1, \theta_2)$ and $b(\theta_1, \theta_2)$. The set Ω need not be an interval; it can be any two-dimensional set.

In this subcase a necessary and sufficient condition for the existence of a minimal pair of jointly sufficient statistics for (θ_1, θ_2) for all $(\theta_1, \theta_2) \in \Omega$ and for all random samples of any size $n \geq 3$ is that $f(x; \theta_1, \theta_2)$ is of the form

$$f(x;\ \theta_1,\ \theta_2) = g(x)\ h(\theta_1,\ \theta_2) \tag{2-10}$$

for all $x \in R$ and for all $(\theta_1, \theta_2) \in \Omega$. When the condition (2-10) is satisfied, the pair $[x_{(1)}, x_{(n)}]$ is jointly minimal sufficient for (θ_1, θ_2), where $x_{(1)}$ and $x_{(n)}$ are respectively the smallest and the greatest members in the sample.

Since both members of the minimal pair of sufficient statistics are functions of only $[x_{(1)}, x_{(n)}]$, the nonregular subcase in point belongs to the pure type.

SUBCASE 2: The ranges (a, b) depend on the single parameter θ_1

Assumptions. Let $f(x; \theta_1, \theta_2)$ be the p.d.f. of the family of distributions. Let the range of admissible values of (θ_1, θ_2) be some two-dimensional nondegenerate interval Ω which may be closed, open, half-open, finite, or infinite. Let Ω_1 be the one-dimensional nondegenerate interval of admissible values of θ_1. Let the set R be the union of the ranges $[a(\theta_1), b(\theta_1)]$ for all $\theta_1 \in \Omega_1$. For all $\theta_1 \in \Omega_1$, $a(\theta_1)$ and $b(\theta_1)$ are continuous, differentiable, and monotonic functions of θ_1, and $a(\theta_1) < b(\theta_1)$. In addition, we assume the Koopman regularity conditions for the general form (2-8) for the regular case with $p = 2$ and $\nu = 1$.

Under these assumptions, for the existence of a minimal pair of jointly sufficient statistics for (θ_1, θ_2) for all $(\theta_1, \theta_2) \in \Omega$ for all random samples of any size $n \geq 3$, it is both necessary and sufficient that (1) $a(\theta_1)$ and $b(\theta_1)$ are monotonic in θ_1 in opposite senses, and that (2) for all $x \in R$ and all $(\theta_1, \theta_2) \in \Omega$, $f(x; \theta_1, \theta_2)$ is of the form

$$f(x;\ \theta_1,\ \theta_2) = \exp[u(\theta_1,\ \theta_2)\ v(x) + A(x) + B(\theta_1,\ \theta_2)]$$
$$[a(\theta_1) \leq x \leq b(\theta_1)], \tag{2-11}$$

where $v(x)$ is not a constant.

When conditions (1) and (2) are satisfied, one member T_1 of the minimal pair (T_1, T_2) of jointly sufficient statistics for (θ_1, θ_2) is a function of $[x_{(1)}, x_{(n)}]$, and the other member T_2

is $\sum_{i=1}^{n} v(x_i)$. For a general formula for T_1 covering different cases, see Huzurbazar (1964). This nonregular base belongs to the mixed type, since one member T_2 of the minimal pair of sufficient statistics is not a function of $[x_{(1)}, x_{(n)}]$.

2.6 Statements of the Solutions of Some Generalized Cauchy Functional Equations

In this section it will be assumed that f(x) is a differentiable function of x defined for all real values of x. We shall now state the general solutions of some generalized Cauchy functional equations satisfied by f(x).

If for all (x, y), f(x) satisfies the functional equation

$$f(x + y) = \phi_1(x)\ \phi_2(y), \tag{2-12}$$

where $\phi_1(x)$ is some function of x and $\phi_2(y)$ is some function of y, then the general solution (Pexider, 1902) of f(x) is

$$f(x) = c_1 \exp(c_2 x), \tag{2-13}$$

where c_1 and c_2 are arbitrary constants. Equation (2-13) is an extension of one of the four well-known Cauchy functional equations in which $f(x) = \phi_1(x) = \phi_2(x)$.

If for all (x, y), f(x) satisfies the functional equation

$$f(xy) = \phi_1(x)\ \phi_2(y), \tag{2-14}$$

where $\phi_1(x)$ is some function of x and $\phi_2(y)$ is some function of y, then the general solution (Pexider, 1902) of f(x) is

$$f(x) = c_1\, x^{c_2}, \tag{2-15}$$

where c_1 and c_2 are arbitrary constants. Equation (2-14) is an extension of another of the four Cauchy functional equations in which $f(x) = \phi_1(x) = \phi_2(x)$. As in the case of the solutions of the Cauchy functional equations, the general solution (2-15) of Eq. (2-14) is also obtained by the substitution of $\log|x|$ for x in the general solution (2-13) of Eq. (2-12).

If for all (x, y), f(x) satisfies the function equation

$$f(x + y) = \sum_{i=1}^{n} \phi_i(x)\ \Psi_i(y), \tag{2-16}$$

where the $\phi_i(x)$ are some functions of x and the $\Psi_i(y)$ are some functions of y, then the general solution (Levi-Civita, 1913; Stäckel, 1913) of f(x) is

$$f(x) = \sum_{i=1}^{n} P_i(x)\ \exp(c_i x), \tag{2-17}$$

where each $P_i(x)$ $(i = 1, \ldots, n)$ is a polynomial in x, and the c_i are constants. Equation (2-16) may be called a generalized Cauchy functional equation.

A particular case. In this discussion we shall have occasion to consider the particular equation

$$f(x + y) = \phi_1(x)\ \Psi_1(y) + \phi_2(x)\ \Psi_2(y), \tag{2-18}$$

obtained by putting n = 2 in Eq. (2-16).

Following the methods of Levi-Civita and Stäckel, it can be shown that in this case the general solution (2-17) takes either the form

$$f(x) = c_1 \exp(c_2 x) + c_3 \exp(c_4 x) \tag{2-19}$$

or $$f(x) = (k_1 x + k_2) \exp(k_3 x), \tag{2-20}$$

where the c's and the k's are arbitrary constants.

We shall also have occasion to consider the functional equation

$$f(xy) = \phi_1(x)\ \Psi_1(y) + \phi_2(x)\ \Psi_2(y), \tag{2-21}$$

analogous to Eq. (2-18). It can be shown that the general solution of Eq. (2-21) is either

$$f(x) = c_1 x^{c_2} + c_3 x^{c_4} \tag{2-22}$$

or $$f(x) = k_1 \log|x| + k_2) x^{k_3}, \tag{2-23}$$

where the c's and the k's are arbitrary constants. It may be noted that Eqs. (2-22) and (2-23) are also obtained by putting $\log |x|$ for x in Eqs. (2-19) and (2-20) respectively.

2.7 A New Functional Equation

Let f(x) be a differentiable function of x defined for all real values of x. If for all (x, y, z), f(x) satisfies the functional equation

$$f(xz + yz) = \phi_1(y, z)\ \Psi_1(x) + \phi_2(y, z)\ \Psi_2(x), \tag{2-24}$$

where ϕ_1 and ϕ_2 are some functions of (y, z), and Ψ_1 and Ψ_2 are some functions of x, then the general solution of f(x) is

$$f(x) = \lambda_1 x + \lambda_2, \tag{2-25}$$

where λ_1 and λ_2 are arbitrary constants.

Proof. Let f(x) satisfy the functional equation (2-24).

Since Eq. (2-24) holds good for all (x, y, z), we may first put in particular z = 1 in Eq. (2-24). Then f(x) satisfies the functional equation

$$f(x + y) = \phi_1(y, 1)\ \Psi_1(x) + \phi_2(y, 1)\ \Psi_2(x),$$

i.e., (2-26)

$$f(x + y) = \phi_{11}(y)\ \Psi_1(x) + \phi_{21}(y)\ \Psi_2(x),$$

which is the same as the functional equation (2-18). Hence, using Eqs. (2-19) and (2-20), the general solution of Eq. (2-26) is either

$$f(x) = c_1 \exp(c_2 x) + c_3\ e(c_4 x) \tag{2-27}$$

or

$$f(x) = (k_1 x + k_2) \exp(k_3 x), \tag{2-28}$$

where the c's and k's are arbitrary constants.

Hence every solution of Eq. (2-24) must satisfy either Eq. (2-27) or Eq. (2-28).

Next put in particular y = 0 in Eq. (2-24). Then f(x) satisfies the functional equation

$$f(xz) = \phi_1(0, z)\ \Psi_1(x) + \phi_2(0, z)\ \Psi_2(x),$$

i.e.,

$$f(xz) = \phi_{12}(z)\ \Psi_1(x) + \phi_{22}(z)\ \Psi_2(x), \tag{2-29}$$

which is the same as the functional equation (2-21). Hence, using Eqs. (2-22) and (2-23), the general solution of Eq. (2-29) is either

$$f(x) = \mu_1 x^{\mu_2} + \mu_3 x^{\mu_4}, \tag{2-30}$$

or $$f(x) = (\nu_1 \log|x| + \nu_2) x^{\nu_3}, \tag{2-31}$$

where the μ's and ν's are arbitrary constants.

Hence every solution of Eq. (2-24) must satisfy either Eq. (2-30) or Eq. (2-31).

Since every solution f(x) of Eq. (2-24) must satisfy one of the pair of equations [(2-27), (2-28)] and also one of the pair of equations [(2-30), (2-31)], it follows from inspection of the two pairs of equations that f(x) must satisfy Eq. (2-25), where λ_1 and λ_2 are arbitrary constants. It may be noted that Eq. (2-28) reduces to Eq. (2-25) if $k_1 = \lambda_1$, $k_2 = \lambda_2$ and $k_3 = 0$; and Eq. (2-30) reduces to Eq. (2-25) if $\mu_1 = \lambda_1$, $\mu_2 = 1$, $\mu_3 = \lambda_2$, and $\mu_4 = 0$.

Conversely, if f(x) satisfies Eq. (2-25), it is easily verified that f(x) satisfies Eq. (2-24). Hence Eq. (2-25) is the general solution of the functional equation (2-24).

A particular case. A particular case of the functional equation (2-24) is obtained by taking either $\phi_2 = 0$ or $\Psi_2 = 0$. Then Eq. (2-24) becomes

$$f(xz + yz) = \phi(y, z)\ \Psi(x). \tag{2-32}$$

Since Eq. (2-32) is a particular case of Eq. (2-24), it follows that every solution f(x) of Eq. (2-32) must satisfy Eq. (2-25). Now it is easily verified that if f(x) satisfies Eq. (2-25), then f(x) satisfies Eq. (2-32) if, and only if, $\lambda_1 = 0$, in which case Eq. (2-25) reduces to $f(x) = \lambda_2 = c$. Hence the general solution of

Eq. (2-32) is

$$f(x) = c, \qquad (2\text{-}33)$$

where c is an arbitrary constant.

3. Location Parameter

3.1 Formal Definition of a Location Parameter

Let $F(x;\ \theta)$ be the d.f. of a family of one dimensional distributions depending on a parameter θ. Let Ω be the set of admissible values of θ. We shall call θ the "location parameter" of the family of distributions if, for all $x \in R_1 = (-\infty, \infty)$ and for all $\theta \in \Omega$, $F(x;\ \theta)$ is of the form

$$F(x;\ \theta) = F_1(x - \theta), \qquad (3\text{-}1)$$

where $F_1(x - \theta)$ is a function of $x - \theta$.

The definition (3-1) holds good for both continuous and discrete distributions, as well as for distributions of both regular and nonregular types as defined in Sec. 2.2. If the family of distributions represented by $F(x;\ \theta)$, possesses a p.d.f. $f(x;\ \theta)$, the relation (3-1) can be written in the form

$$f(x;\ \theta) = f_1(x - \theta), \qquad (3\text{-}2)$$

where $f_1(x - \theta)$ is a function of $x - \theta$.

We shall use the phrase "a location-parameter family of distributions" if the family of distributions depends on a location parameter.

3.2 The Range of a Location Parameter

We note that if any function of the type $F_1(x - \theta)$ satisfies the conditions for being the d.f. for some value, say θ_0, of θ, then

$$F_1(x - \theta) = F_1[(x - \theta_0) + \theta_0 - \theta]$$

satisfies the conditions for being the d.f. for every $\theta \in R_1$.

Hence the natural range of values of a location parameter θ is the interval $\Omega = R_1 = (-\infty, \infty)$.

3.3 The Reduced Form of a Location-Parameter Family of Distributions

The distribution corresponding to a selected value of the location parameter θ, say $\theta = 0$, will be called the "reduced form" of the location-parameter family of distributions. For the location-parameter family of distributions represented by the d.f. $F_1(x - \theta)$, the d.f. of the reduced form is $F_1(x)$. It may be noted that $F_1(x)$ is independent of θ, and the general d.f. $F_1(x - \theta)$ of the location-parameter family of distributions is obtained by writing $x - \theta$ for x in the reduced d.f. $F_1(x)$. $F_1(x)$ is the d.f. of the random variable $X - \theta$. It also follows that every distribution with d.f. $F_1(x)$ (not depending on θ) is a member of the location-parameter family of distributions with d.f. $F_1(x - \theta)$.

3.4 The Range of a Location-parameter Family of Distributions

The range (a_0, b_0) (as defined in Sec. 2.1) of the reduced form $F_1(x)$ will be called the "reduced range." It follows that a_0 and b_0 are independent of θ. The ranges $[a(\theta), b(\theta)]$ of the family of distributions $F_1(x - \theta)$ are then given by

$$a(\theta) = a_0 + \theta \tag{3-3}$$

and $$b(\theta) = b_0 + \theta. \tag{3-4}$$

From Eqs. (3-3), and (3-4) it follows that the ranges $[a(\theta), b(\theta)]$ of a location-parameter family of distributions will be independent of θ if, and only if,

$$a_0 = -\infty, \qquad b_0 = +\infty, \tag{3-5}$$

in which case $R_1 = (-\infty, \infty)$ is the range of the location-parameter family of distributions. We have thus proved that the range of a location-parameter family of distributions of the regular type is R_1, which is also the range of the location parameter.

For a location-parameter family of distributions of the nonregular type, the terminals of the ranges are given by Eqs. (3-3) and (3-4), where at least one of a_0 and b_0 is finite. It may be noted from Eqs. (3-3) and (3-4) that each of the terminals differs from the location parameter by an additive constant only. If both terminals of the ranges depend on θ, the ranges are $(a_0 + \theta, b_0 + \theta)$, where both a_0 and b_0 are finite. If only one of the two terminals depends on θ, there will be two subcases:

1. $a(\theta)$ depends on θ, but $b(\theta)$ is independent of θ. This subcase arises if, and only if, a_0 is finite and $b_0 = +\infty$. Then $a(\theta) = a_0 + \theta$ and $b(\theta) = +\infty$. The ranges in this subcase are therefore $(a_0 + \theta, \infty)$.

2. $a(\theta)$ is independent of θ, but $b(\theta)$ depends on θ. In this subcase we have $a_0 = -\infty$ and b_0 is finite, so that $a = -\infty$ and $b(\theta) = b_0 + \theta$. Hence the ranges are $(-\infty, b_0 + \theta)$.

We have thus shown that for a location-parameter family of distributions of the nonregular type, the ranges belong to one of the following three types:

$$\text{Type 1:} \quad (a_0 + \theta, b_0 + \theta), \tag{3-6}$$

$$\text{Type 2:} \quad (-\infty, b_0 + \theta), \tag{3-7}$$

$$\text{Type 3:} \quad (a_0 + \theta, \infty), \tag{3-8}$$

where a_0 and b_0 are finite.

3.5 A Property of the Location Parameter

If a family of distributions depends on a location parameter, then in reparameterization the choice of the location parameter is unique except for an arbitrary additive constant.

Proof. If θ is the location parameter of a family of distributions with d.f. $F(x; \theta)$, we have

$$F(x; \theta) = F_1(x - \theta). \tag{3-9}$$

First consider the transformation $\theta \leftrightarrow \alpha$ such that

$$\theta = \alpha + k, \tag{3-10}$$

where k is a constant. Let the d.f. $F(x;\ \theta)$ transform into the d.f. $\phi(x;\ \alpha)$, where

$$\begin{aligned}\phi(x;\ \alpha) &= F(x;\ \alpha + k)\\ &= F_1(x - \alpha - k) \qquad \text{[in virtue of Eq. (3-9)]}\\ &= \phi_1(x - \alpha),\end{aligned} \tag{3-11}$$

where $\phi_1(x) = F_1(x - k)$. (3-12)

Equation (3-11) shows that α is also the location parameter of the given family of distributions.

Conversely, suppose that both θ and $\alpha = \Psi(\theta)$ are location parameters of the same family of distributions. Let the d.f. of the family be $F_1(x - \theta)$ when expressed in terms of θ, and let the d.f. be $\phi_1(x - \alpha)$ when expressed in terms of α. Then for all x and for all corresponding values of θ and α, we have

$$F_1(x - \theta) = \phi_1(x - \alpha),$$

i.e.,

$$F_1(x - \theta) = \phi_1[x - \Psi(\theta)]. \tag{3-13}$$

We assume that $F_1(x)$ is a differentiable function of x, which is equivalent to the assumption that the given location-parameter family of distributions possesses a p.d.f. Differentiating Eq. (3-13) with respect to x, we have

$$F_1'(x - \theta) = \phi_1'[x - \Psi(\theta)], \tag{3-14}$$

where the prime denotes differentiation.

Again differentiating Eq. (3-13) with respect to θ, we have

$$F_1'(x - \theta) = \phi_1'[x - \Psi(\theta)]\ \Psi'(\theta). \tag{3-15}$$

From Eqs. (3-14) and (3-15) we have $\Psi'(\theta) = 1$, which gives

$$\Psi(\theta) = \theta + k,$$

i.e., $\alpha = \theta + k,$ (3-16)

where k is an arbitrary constant. Hence the property is proved.

3.6 A Property of the Variate of a Location-Parameter Family of Distributions

A property analogous to the property of Sec. 3.5 holds good also for the variate of the location-parameter family of distributions:

Let X be the variate of a location-parameter family of distributions, and consider any one-to-one transformation $X \leftrightarrow Y$. Then Y will be the variate of a location-parameter family of distributions if, and only if,

$$Y = X + k, \qquad (3\text{-}17)$$

where k is an arbitrary constant.

The proof of the property (3-17) is exactly on similar lines as the proof of the property in Sec. 3.5.

3.7 A Property of the Location-Parameter Family of Distributions

The properties of Sections 3.5 and 3.6 may be combined as follows:

In a variate transformation or a parameter transformation or both, a location-parameter family of distributions will continue to be a location-parameter family of distributions if, and only if, each transformation is linear with change of origin but without change of scale.

4. Scale Parameter

4.1 Formal Definition of a Scale Parameter

Let $F(x;\ \sigma)$ be the d.f. of a family of one-dimensional distributions depending on a positive parameter σ. Let Ω be the set of admissible values of σ. We shall call σ the "scale parameter" of the family of distributions if, for all $x \in R_1$ and for all $\sigma \in \Omega$, $F(x;\ \sigma)$ is of the form

$$F(x;\ \sigma) = F_1\left(\frac{x}{\sigma}\right), \qquad (4\text{-}1)$$

The definition (4-1) holds good for both continuous and discrete distributions, as well as for distributions of both regular and nonregular types. If the family of distributions represented by $F(x;\ \sigma)$ possesses a p.d.f. $f(x;\ \sigma)$, the relation (4-1) can be written in the form

$$f(x;\ \sigma) = \frac{1}{\sigma} f_1\left(\frac{x}{\sigma}\right), \tag{4-2}$$

where $f_1(x/\sigma)$ is a function of x/σ.

We shall use the phrase "a scale-parameter family of distributions" if the family of distributions depends on a scale parameter.

4.2 The Range of a Scale Parameter

We note that if any function of the type $F_1(x/\sigma)$ satisfies the conditions for being the d.f. for some positive value, say σ_0, of σ, then

$$F_1\left(\frac{x}{\sigma}\right) = F_1\left(\frac{x}{\sigma_0}\,\frac{\sigma_0}{\sigma}\right)$$

satisfies the conditions for being the d.f. for every positive value of σ. Hence the natural range of values of a scale parameter σ is the open interval $\Omega = (0, \infty)$.

4.3 The Reduced Form of a Scale-Parameter Family of Distributions

The distribution corresponding to a selected value of the scale parameter σ, say $\sigma = 1$, will be called the "reduced form" of the scale-parameter family of distributions. For the scale-parameter family of distributions represented by the d.f. $F_1(x/\sigma)$, the d.f. of the reduced form is $F_1(x)$. It may be noted that the general d.f. $F_1(x/\sigma)$ of the scale-parameter family of distributions is obtained by writing x/σ for x in the reduced d.f. $F_1(x)$. $F_1(x)$ is the d.f. of the random variable x/σ. It also follows that every distribution with d.f. $F_1(x)$ (not depending on σ) is a member of the scale-parameter family of distributions with d.f. $F_1(x/\sigma)$.

4.4 The Range of a Scale-Parameter Family of Distributions

The range (a_0, b_0) of the reduced form $F_1(x)$ will be called the "reduced range." a_0 and b_0 are independent of σ. The ranges $[a(\sigma), b(\sigma)]$ of the family of distributions $F_1(x/\sigma)$ are then given by

$$a(\sigma) = a_0\sigma \tag{4-3}$$

and $$b(\sigma) = b_0\sigma. \tag{4-4}$$

From Eqs. (4-3) and (4-4) it follows that the ranges $[a(\sigma), b(\sigma)]$ of a scale parameter family of distributions will be independent of σ if, and only if, one of the following three cases holds good:

1. $a_0 = -\infty$ and $b_0 = +\infty$, in which case the range of the scale-parameter family of distributions is $(-\infty, \infty)$.
2. $a_0 = -\infty$ and $b_0 = 0$, in which case the range is $(-\infty, 0)$.
3. $a_0 = 0$ and $b_0 = +\infty$, in which case the range is $(0, \infty)$.

We have thus proved that the range of a scale-parameter family of distributions of the regular type is either $(-\infty, \infty)$ or $(-\infty, 0)$ or $(0, \infty)$.

For a scale-parameter family of distributions of the nonregular type, the terminals of the ranges are given by Eqs. (4-3) and (4-4), where the pair (a_0, b_0) does not assume any of the values in cases 1, 2, and 3 which give regular-type distributions. It may be noted from Eqs. (4-3) and (4-4) that each of the terminals is a multiple of the scale parameter. For a scale-parameter family of distributions of the nonregular type, the ranges will be of five different types corresponding to different values of a_0 and b_0 in the following five subcases:

1. $a_0 \neq -\infty$ or 0; $b_0 \neq 0$ or ∞. (4-5)

In this subcase the ranges are $(a_0\sigma, b_0\sigma)$.

2. $a_0 = -\infty$; $b_0 \neq 0$ or ∞. (4-6)

In this subcase the ranges are $(-\infty, b_0\sigma)$.

3. $a_0 = 0$; $b_0 \neq 0$ or ∞. (4-7)

In this subcase $b_0 > 0$ since $b_0 > a_0$, and the ranges are $(0, b_0\sigma)$.

4. $a_0 \neq -\infty$ or 0; $b_0 = \infty$. (4-8)

In this subcase the ranges are $(a_0\sigma, \infty)$.

5. $a_0 \neq -\infty$ or 0; $b_0 = 0$. (4-9)

In this subcase $a_0 < 0$ since $a_0 < b_0$, and the ranges are $(a_0\sigma, 0)$.

It may be noted that in Eq. (4-5) boththe terminals of the ranges depend on σ. In Eqs. (4-6) and (4-7) only the second terminal depend on σ. In subcases (4-8) and (4-9) only the first terminal depends on σ.

4.5 A Property of the Scale Parameter

If a family of distributions depends on a scale parameter, then in reparameterization the choice of the scale parameter is unique except for a positive multiplicative constant.

Proof. If σ is the scale parameter of a family of distributions with d.f. $F(x; \sigma)$, we have

$$F(x; \sigma) = F_1\left(\frac{x}{\sigma}\right). \quad (4\text{-}10)$$

First consider the transformation $\sigma \leftrightarrow \beta$ such that

$$\sigma = k\beta, \quad (4\text{-}11)$$

where k is a positive constant. Let the d.f. $F(x; \sigma)$ transform into the d.f. $\phi(x; \beta)$, where

$$\begin{aligned}\phi(x; \beta) &= F(x; k\beta) \\ &= F_1\left(\frac{x}{k\beta}\right) \quad \text{[in virtue of Eq. (4-10)]} \\ &= \phi_1\left(\frac{x}{\beta}\right), \quad (4\text{-}12)\end{aligned}$$

where $\phi_1(x) = F_1\left(\frac{x}{k}\right)$. (4-13)

Equation (4-12) shows that β is also the scale parameter of the given family of distributions.

Conversely, suppose that both σ and $\beta = \Psi(\sigma)$ are scale parameters of the same family of distributions. Let the d.f. of the family be $F_1(x/\sigma)$ when expressed in terms of σ, and let the d.f. be $\phi_1(x/\beta)$ when expressed in terms of β. Then for all x and for all corresponding values of σ and β we have

$$F_1\left(\frac{x}{\sigma}\right) = \phi_1\left(\frac{x}{\beta}\right),$$

i.e.,

$$F_1\left(\frac{x}{\sigma}\right) = \phi_1\left[\frac{x}{\Psi(\sigma)}\right]. \tag{4-14}$$

We assume that $F_1(x)$ is a differentiable function of x. Differentiating Eq. (4-14) with respect to x, we have

$$\frac{1}{\sigma} F_1'\left(\frac{x}{\sigma}\right) = \frac{1}{\Psi(\sigma)} \phi_1'\left[\frac{x}{\Psi(\sigma)}\right]. \tag{4-15}$$

Again differentiating Eq. (4-14) with respect to σ,

$$\frac{1}{\sigma^2} F_1'\left(\frac{x}{\sigma}\right) = \frac{\Psi'(\sigma)}{[\Psi(\sigma)]^2} \phi_1'\left[\frac{x}{\Psi(\sigma)}\right]. \tag{4-16}$$

From Eqs. (4-15) and (4-16) we have

$$\frac{\Psi'(\sigma)}{\Psi(\sigma)} = \frac{1}{\sigma},$$

which gives $\Psi(\sigma) = k\sigma$, i.e.,

$$\beta = k\sigma, \tag{4-17}$$

where k is an arbitrary positive constant. Hence the property is proved.

4.6 A property of the Variate of a Scale-parameter Family of Distributions

Let X be the variate of a scale-parameter family of distributions, and consider any one-to-one transformation $X \leftrightarrow Y$. Then Y will be the variate of a scale-parameter family of distributions if, and only if,

$$Y = kX, \tag{4-18}$$

where k is a nonzero arbitrary constant. The proof of the propetty (4-18) is exactly on similar lines as the proof of the property in Sec. 4.5.

4.7 A Property of the Scale-parameter Family of Distributions

The properties of Secs. 4.5 and 4.6 may be combined as follows:

In a variate transformation or a parameter transformation or both, a scale-parameter family of distributions will continue to be a scale-parameter family of distributions if, and only if, each transformation is linear with change of scale but without change of origin.

4.8 Relation Between the Location and Scale Parameters

It is well known that if θ is the location parameter of the distribution of X, then $\sigma = e^{\theta}$ is a scale parameter of the distribution of $Y = e^{X}$. This relation, however, unnecessarily restricts the range of the distribution of Y to the interval $(0, \infty)$ in the regular case, and to a part of the interval $(0, \infty)$ in the nonregular case.

As we have already observed in Sec. 4.4, in the general case the ranges of a scale-parameter family of distributions can be intervals different from $(0, \infty)$.

5. Distributions Depending on Both Location and Scale Parameters

5.1 Joint Occurrence of Location and Scale Parameters

If a family of distributions depends on a location parameter θ and also on a scale parameter σ, the d.f. $F(x;\ \theta,\ \sigma)$ is of the form

$$F(x;\ \theta,\ \sigma) = F_1\left(\frac{x - \theta}{\sigma}\right), \tag{5-1}$$

where $F_1[(x - \theta)/\sigma]$ is a function of $(x - \theta)/\sigma$.

If the family of distributions possesses a p.d.f. $f[(x - \theta)/\sigma]$, the relation (5-1) can be written in the form

$$f(x;\ \theta,\ \sigma) = \frac{1}{\sigma} f_1 \frac{x - \theta}{\sigma}, \qquad (5\text{-}2)$$

where $f_1[(x - \theta)/\sigma]$ is a function of $(x - \theta)/\sigma$.

If the parameters of a two-parameter family of distributions are of location and scale, we shall call the family "a location-scale-parameter family of distributions."

5.2 The Reduced Form of a Location-Scale-parameter Family of Distributions

The distribution corresponding to $\theta = 0$ and $\sigma = 1$ will be called the "reduced form" of a location-scale-parameter family of distributions. For a location-scale-parameter family of distributions represented by the d.f. $F_1[(x - \theta)/\sigma]$, the d.f. of the reduced form is $F_1(x)$. It may be noted that $F_1(x)$ is the d.f. of the random variable $(X - \theta)/\sigma$.

5.3 The Range of a Location-Scale-parameter Family of Distributions

The range (a_0, b_0) of the reduced form $F_1(x)$ will be called the "reduced range." The ranges $[a(\theta, \sigma), b(\theta, \sigma)]$ of the family of distributions $F_1[(x - \theta)/\sigma]$ are then given by

$$a(\theta, \sigma) = \theta + a_0\sigma \qquad (5\text{-}3)$$

and $$b(\theta, \sigma) = \theta + b_0\sigma. \qquad (5\text{-}4)$$

From Eqs. (5-3) and (5-4) it follows that the ranges $[a(\theta, \sigma), b(\theta, \sigma)]$ will be independent of (θ, σ) if, and only if,

$$a_0 = -\infty \quad \text{and} \quad b_0 = +\infty, \qquad (5\text{-}5)$$

in which case R_1 is the range of the location-scale-parameter family of distributions. We have thus shown that the range of a location-scale-parameter family of distributions of the regular type is R_1.

For a location-scale-parameter family of distributions of the nonregular type, the terminals of the ranges are given by Eqs. (5-3) and (5-4), where at least one of a_0 and b_0 is finite. It may be noted from Eqs. (5-3) and (5-4) that each of the terminals is equal to the sum of the location parameter and a constant multiple of the scale parameter. The ranges of a location-scale-parameter family of distributions of the nonregular type will be of seven different types corresponding to different values of a_0 and b_0 in the following seven subcases:

1. $a_0 \neq -\infty$ or 0, $b_0 \neq 0$ or ∞. (5-6)

In this subcase the ranges are $(\theta + a_0\sigma, \theta + b_0\sigma)$.

2. $a_0 = 0$, $b_0 \neq 0$ or ∞. Then $b_0 > 0$. (5-7)

The ranges are $(\theta, \theta + b_0\sigma)$.

3. $a_0 \neq -\infty$ or 0, $b_0 = 0$. Then $a_0 < 0$. (5-8)

The ranges are $(\theta + a_0\sigma, \theta)$.

4. $a_0 = -\infty$, $b_0 \neq 0$ or ∞. (5-9)

The ranges are $(-\infty, \theta + b_0\sigma)$.

5. $a_0 \neq -\infty$ or 0, $b_0 = \infty$. (5-10)

The ranges are $(\theta + a_0\sigma, \infty)$.

6. $a_0 = -\infty$, $b_0 = 0$. (5-11)

The ranges are $(-\infty, \theta)$.

7. $a_0 = 0$, $b_0 = \infty$. (5-12)

The ranges are (θ, ∞).

It may be noted that in Eqs. (5-6) - (5-12) the ranges depend on both parameters θ and σ. In Eq. (5-6) each terminal depends on both parameters θ and σ. In Eqs. (5-7) and (5-8) one terminal depends on both parameters, and the other depends on only one of the two parameters. In Eqs. (5-9) and (5-10) one terminal depends on both parameters, but the other terminal is fixed (independent of the parameters). In Eqs. (5-11) and (5-12) the range depends on only one of the two parameters.

It is interesting to note in Eqs. (5-11) and (5-12) that if the range depends on a single parameter only, the parameter must be that of location, i.e., we do not get a subcase where the range depends on the scale parameter only. It may also be noted that in Eqs. (5-11) and (5-12), only one terminal is the parameter of location; the other terminal is either $-\infty$ or $+\infty$. In Eqs. (5-7) and (5-8) the terminal depending on a single parameter is equal to the location parameter itself.

5.4 A Joint Property of Location and Scale Parameters

If a family of distributions depends on both location and scale parameters, and if (θ_1, σ_1) and (θ_2, σ_2) are pairs of location and scale parameters corresponding to two different ways of reparameterization, then

$$\theta_1 = \theta_2 + \lambda_1\sigma_2 \tag{5-13}$$

and $$\sigma_1 = \lambda_2\sigma_2, \tag{5-14}$$

where λ_1 and λ_2 $(\lambda_2 > 0)$ are arbitrary constants.

Proof. First we show that if (θ_1, σ_1) is a pair of location and scale parameters for a family of distributions, then under the reparameterization transformation (5-13) and (5-14), (θ_2, σ_2) is also a pair of location and scale parameters for the same family of distributions. For if $F(x; \theta_1, \sigma_1)$ is the d.f. of the family of distributions, we have

$$F(x; \theta_1, \sigma_1) = F_1\left(\frac{x - \theta_1}{\sigma_1}\right). \tag{5-15}$$

Let the d.f. $F(x;\ \theta_1,\ \sigma_1)$ transform into the d.f. $\phi(x;\ \theta_2,\ \sigma_2)$, where

$$\phi(x;\ \theta_2,\ \sigma_2) = F(x;\ \theta_2 + \lambda_1\sigma_2,\ \lambda_2\sigma_2)$$

$$= F_1\left(\frac{x - \theta_2 - \lambda_1\sigma_2}{\lambda_2\sigma_2}\right) \qquad \text{[in virtue of Eqs. (5-13) and (5-14)]}$$

$$= F_1\left(\frac{1}{\lambda_2}\,\frac{x - \theta_2}{\sigma_2} - \frac{\lambda_1}{\lambda_2}\right)$$

$$= \phi_1\left(\frac{x - \theta_2}{\sigma_2}\right), \tag{5-16}$$

where

$$\phi_1(x) = F_1\left(\frac{x}{\lambda_2} - \frac{\lambda_1}{\lambda_2}\right). \tag{5-17}$$

Equation (5-16) shows that $(\theta_2,\ \sigma_2)$ is a pair of location and scale parameters for the same family of distributions.

Next suppose that $(\theta_1,\ \sigma_1)$ and $(\theta_2,\ \sigma_2)$ are pairs of location and scale parameters corresponding to two different ways of reparameterization for the same family of distributions, so that we may regard each of θ_1 and σ_1 as a function of $(\theta_2,\ \sigma_2)$ and vice versa. If $F(x;\ \theta_1,\ \sigma_1)$ and $\phi(x;\ \theta_2,\ \sigma_2)$ are the d.f.'s corresponding to the two ways of reparameterization, we have

$$F(x;\ \theta_1,\ \sigma_1) = \phi(x;\ \theta_2,\ \sigma_2) \tag{5-18}$$

for all x and for all corresponding values of the two pairs $(\theta_1,\ \sigma_1)$ and $(\theta_2,\ \sigma_2)$ of parameters.

Also, since

$$F(x;\ \theta_1,\ \sigma_1) = F_1\left(\frac{x - \theta_1}{\sigma_1}\right) \tag{5-19}$$

and

$$\phi(x;\ \theta_2,\ \sigma_2) = \phi_1\left(\frac{x - \theta_2}{\sigma_2}\right), \tag{5-20}$$

we have

$$F_1\left(\frac{x - \theta_1}{\sigma_1}\right) = \phi_1\left(\frac{x - \theta_2}{\sigma_2}\right), \tag{5-21}$$

for all x, and for all corresponding values of (θ_1, σ_1) and (θ_2, σ_2). In Eq. (5-21) write for convenience

$$\alpha_1 = -\theta_1, \quad \beta_1 = \frac{1}{\sigma_1}, \tag{5-22}$$

and $$\alpha_2 = -\theta_2, \quad \beta_2 = \frac{1}{\sigma_2}. \tag{5-23}$$

Then Eq. (5-21) becomes

$$F_1(\beta_1 x + \alpha_1\beta_1) = \phi_1(\beta_2 x + \alpha_2\beta_2). \tag{5-24}$$

We assume that the given family of distributions possesses a p.d.f., so that $F_1(x)$ and $\phi_1(x)$ are differentiable functions of x. Differentiating Eq. (5-24) with respect to x,

$$\beta_1 F_1'(\beta_1 x + \alpha_1\beta_1) = \beta_2\ \phi_1'(\beta_2 x + \alpha_2\beta_2). \tag{5-25}$$

Differentiating Eq. (5-24) with respect to α_2 (and noting that each of α_1 and β_1 is a function of both α_2 and β_2),

$$\left[\frac{\partial\beta_1}{\partial\alpha_2} x + \frac{\partial}{\partial\alpha_2}(\alpha_1\beta_1)\right] F_1'(\beta_1 x + \alpha_1\beta_1) = \beta_2\ \phi_1'(\beta_2 x + \alpha_2\beta_2). \tag{5-26}$$

Differentiating Eq. (5-24) with respect to β_2,

$$\left[\frac{\partial\beta_1}{\partial\beta_2} x + \frac{\partial}{\partial\beta_2}(\alpha_1\beta_1)\right] F_1'(\beta_1 x + \alpha_1\beta_1) = (x + \alpha_2)\ \phi_1'(\beta_2 x + \alpha_2\beta_2). \tag{5-27}$$

From Eqs. (5-25) and (5-27) we have

$$\frac{1}{\beta_1}\frac{\partial\beta_1}{\partial\beta_2} x + \frac{1}{\beta_1}\frac{\partial}{\partial\beta_2}(\alpha_1\beta_1) = \frac{1}{\beta_2} x + \frac{\alpha_2}{\beta_2}. \tag{5-28}$$

Since Eq. (5-28) holds good for all x and for all corresponding values of (α_1, β_1) and (α_2, β_2), we have

$$\frac{1}{\beta_1}\frac{\partial\beta_1}{\partial\beta_2} = \frac{1}{\beta_2}, \tag{5-29}$$

$$\text{and } \frac{1}{\beta_1}\frac{\partial}{\partial\beta_2}(\alpha_1\beta_1) = \frac{\alpha_2}{\beta_2}, \tag{5-30}$$

Equation (5-29) gives

$$\beta_1 = \beta_2\, \Psi_1(\alpha_2), \tag{5-31}$$

where $\Psi_1(\alpha_2)$ is an arbitrary function of α_2. From Eqs. (5-25) and (5-26) we have

$$\frac{\partial\beta_1}{\partial\alpha_2} x + \frac{\partial}{\partial\alpha_2}(\alpha_1\beta_1) = \beta_1. \tag{5-32}$$

Since Eq. (5-32) holds good for all x and for all corresponding values of (α_1, β_1) and (α_2, β_2), we have

$$\frac{\partial\beta_1}{\partial\alpha_2} = 0 \tag{5-33}$$

$$\text{and } \frac{\partial}{\partial\alpha_2}(\alpha_1\beta_1) = \beta_1. \tag{5-34}$$

Equation (5-33) gives

$$\beta_1 = \Psi_2(\beta_2), \tag{5-35}$$

where Ψ_2 is an arbitrary function of β_2. From Eqs. (5-31) and (5-35) we must have

$$\beta_1 = k_1\beta_2, \tag{5-36}$$

where k_1 is an arbitrary positive constant. Equations (5-34) and (5-36) give

$$\frac{\partial}{\partial\alpha_2}(k_1\alpha_1\beta_2) = k_1\beta_2,$$

i.e.,

$$k_1\beta_2 \frac{\partial\alpha_1}{\partial\alpha_2} = k_1\beta_2,$$

$$\text{i.e., } \frac{\partial\alpha_1}{\partial\alpha_2} = 1. \tag{5-37}$$

Equation (5-37) gives

$$\alpha_1 = \alpha_2 + \Psi_3(\beta_2), \tag{5-38}$$

where Ψ_3 is an arbitrary function of β_2. Equations (5-30) and (5-36) give

$$\frac{1}{k_1\beta_2}\frac{\partial}{\partial\beta_2}(k_1\alpha_1\beta_2) = \frac{\alpha_2}{\beta_2},$$

i.e.,
$$\frac{\partial}{\partial\beta_2}(\alpha_1\beta_2) = \alpha_2,$$

i.e.,
$$\beta_2\frac{\partial\alpha_1}{\partial\beta_2} + \alpha_1 = \alpha_2. \tag{5-39}$$

Equations (5-38) and (5-39) give

$$\beta_2\,\Psi_3'(\beta_2) + \alpha_2 + \Psi_3(\beta_2) = \alpha_2,$$

i.e.,
$$\beta_2\,\Psi_3'(\beta_2) + \Psi_3(\beta_2) = 0$$

i.e.,
$$\frac{\Psi_3'(\beta_2)}{\Psi_3(\beta_2)} + \frac{1}{\beta_2} = 0. \tag{5-40}$$

Solving Eq. (5-40) for $\Psi_3(\beta_2)$,

$$\Psi_3(\beta_2) = \frac{k_2}{\beta_2}, \tag{5-41}$$

where k_2 is an arbitrary constant. Then Eq. (5-38) gives

$$\alpha_1 = \alpha_2 + \frac{k_2}{\beta_2}. \tag{5-42}$$

Thus Eqs. (5-36) and (5-42) give α_1 and β_1 in terms of α_2 and β_2. Substituting back from Eqs. (5-22) and (5-23) for α_1, β_1, α_2, and β_2 in terms of θ_1, σ_1, θ_2, and σ_2, we have the following:

Equation (5-42) becomes

$$-\theta_1 = -\theta_2 + k_2\sigma_2,$$

i.e.,
$$\theta_1 = \theta_2 + \lambda_1\sigma_2, \tag{5-43}$$

where λ_1 is an arbitrary constant, equation (5-36) becomes

$$\frac{1}{\sigma_1} = \frac{k_1}{\sigma_2},$$

i.e.,

$$\sigma_1 = \lambda_2\sigma_2, \qquad (5\text{-}44)$$

where λ_2 is an arbitrary positive constant.

Thus Eqs. (5-43) and (5-44) show that the relations (5-13) and (5-14) hold good, and hence the property is established.

5.5 A Property of the Variate of a Location-Scale-parameter Family of Distributions

Let X be the variate of a location-scale-parameter family of distributions, and consider any one-to-one transformation $X \leftrightarrow Y$. Then Y will be the variate of a location-scale-parameter family of distributions if, and only if,

$$Y = k_1x + k_2, \qquad (5\text{-}45)$$

where k_1 ($k_1 \neq 0$) and k_2 are arbitrary constants. The proof of the property (5-45) is exactly on similar lines as the proof of the property in Sec. 5.4.

5.6 A Property of the Location-Scale-parameter Family of Distributions

The properties of Secs. 5.4 and 5.5 may be combined as follows: Let the variate X and the parameters (θ, σ) of a location-scale-parameter family of distributions undergo one-to-one transformations:

$$X \leftrightarrow Y, \quad (\theta, \sigma) \leftrightarrow (\alpha, \beta). \qquad (5\text{-}46)$$

Then the transformed family of distributions of Y with parameters (α, β) will be a location-scale-parameter family of distributions if, and only if,

$$y = k_1x + k_2, \quad \alpha = \theta + \lambda_1\sigma, \quad \text{and} \quad \beta = \lambda_2\sigma, \qquad (5\text{-}47)$$

where k_1 ($k_1 \neq 0$), k_2, λ_1, and λ_2 ($\lambda_2 > 0$) are arbitrary constants.

6. Location-Parameter Families of Distributions of the Regular Type Admitting a Single Sufficient Statistic for the Location Parameter

6.1 Statement of the Problem and the Assumptions

Let $f(x; \theta) = f_1(x - \theta)$ be the p.d.f. of a location-parameter family of distributions of the regular type. Since the distributions are of the regular type, it follows from Sec. 3.2 that the range of θ is R_1, and it follows from Sec. 3.4 that R_1 is also the range of the family of distributions. We wish to determine $f(x; \theta)$ so that a single sufficient statistic exists for θ for all $\theta \in R_1$ and for all random samples of any size $n \geq 2$.

Assumptions. We assume the Koopman regularity conditions for the general form (2-6) for existence of a single sufficient statistic for a single unknown parameter θ in the regular case. The Koopman regularity conditions imply mainly the existence of $\partial^2 f/\partial\theta\, \partial x$, which is equivalent to the existence of $f_1''(x)$. We further assume the existence of the third derivative $f_1'''(x)$.

6.2 Solution of the Problem in Section 6.1

Under the assumptions of Sec. 6.1 we have using Eq. (2-6)

$$f(x; \theta) = f_1(x - \theta) = \exp[u(\theta)\, v(x) + A(x) + B(\theta)] \qquad (6\text{-}1)$$

for all $x \in R_1$ and for all $\theta \in R_1$.

Writing $-\theta$ for θ in Eq. (6-1) we have

$$\begin{aligned} f_1(x + \theta) &= \exp[u(-\theta)\, v(x) + A(x) + B(-\theta)] \\ &= \exp[u_1(\theta)\, v(x) + A(x) + B_1(\theta)], \end{aligned} \qquad (6\text{-}2)$$

where $u_1(\theta) = u(-\theta)$ and $B_1(\theta) = B(-\theta)$. We have

$$\log f_1(x + \theta) = u_1(\theta)\, v(x) + A(x) + B_1(\theta). \qquad (6\text{-}3)$$

Write $$f_2(x) = \log f_1(x). \qquad (6\text{-}4)$$

Then Eq. (6-3) becomes

$$f_2(x + \theta) = u_1(\theta)\ v(x) + A(x) + B_1(\theta). \tag{6-5}$$

Differentiating Eq. (6-5) with respect to x,

$$f_2'(x + \theta) = u_1(\theta)\ v'(x) + A'(x). \tag{6-6}$$

Differentiating Eq. (6-6) with respect to θ,

$$f_2''(x + \theta) = u_1'(\theta)\ v'(x). \tag{6-7}$$

Writing $f_3(x) = f_2''(x)$, (6-8)

Eq. (6-7) is of the form

$$f_3(x + \theta) = \phi_1(x)\ \phi_2(\theta), \tag{6-9}$$

where $\phi_1(x)$ is some function of x, and $\phi_2(\theta)$ is some function of θ. Equation (6-9) is a functional equation satisfied by $f_3(x)$ for all $x \in R_1$ and for all $\theta \in R_1$. $f_3'(x)$ exists, since $f''{}_1'(x)$ exists. Hence, using results of Sec. 2.6.1, the general solution of Eq. (6-9) is

$$f_3(x) = c_1 \exp(c_2 x), \tag{6-10}$$

where c_1 and c_2 are arbitrary constants. From Eqs. (6-8) and (6-10) we have

$$f_2''(x) = c_1 \exp(c_2 x). \tag{6-11}$$

Two cases arise:

CASE I: $c_2 \neq 0$.

In this case the solution of Eq. (6-11) is

$$f_2(x) = \frac{c_1}{c_2^2} \exp(c_2 x) + c_3 x + c_4, \tag{6-12}$$

where c_1, c_2 ($c_2 \neq 0$), c_3, and c_4 are arbitrary constants. Without any loss of generality, we may replace c_1/c_2^2 by c_1, so that

$$f_2(x) = c_1 \exp(c_2 x) + c_3 x + c_4 \quad (c_2 \neq 0). \tag{6-13}$$

From Eqs. (6-4) and (6-13) we have

$$f_1(x) = \exp[c_1 \exp(c_2x) + c_3x + c_4] \quad (c_2 \neq 0). \tag{6-14}$$

CASE II: $c_2 = 0$ in Eq. (6-11).

In this case Eq. (6-11) becomes

$$f''_2(x) = c_1, \tag{6-15}$$

which gives

$$f_2(x) = \frac{1}{2} c_1x^2 + c_2x + c_3, \tag{6-16}$$

where c_1, c_2, and c_3 are arbitrary constants. Replacing $\frac{1}{2}c_1$ by c_1, we have

$$f_2(x) = c_1x^2 + c_2x + c_3. \tag{6-17}$$

From Eqs. (6-4) and (6-17) we have

$$f_1(x) = \exp(c_1x^2 + c_2x + c_3). \tag{6-18}$$

We have thus shown that if $f_1(x)$ satisfies the functional equation (6-1), then $f_1(x)$ must satisfy either Eq. (6-14) or Eq. (6-18). Conversely, it is easily verified that if $f_1(x)$ satisfies either Eq. (6-14) or Eq. (6-18), it satisfies Eq. (6-1) also. Hence Eqs. (6-14) and (6-18) constitute the general solution of the functional equation (6-1). Since $f_1(x)$ is the p.d.f. of the reduced form of the family of distributions with p.d.f. $f(x; \theta) = f_1(x - \theta)$, we have shown that Eqs. (6-14) and (6-18) give the p.d.f. of the reduced forms of the location-parameter families of distributions admitting a single sufficient statistic for the location parameter. The arbitrary constants occurring in the reduced forms (6-14) and (6-18) may be called "other parameters," which will be treated as known in estimation of the location parameter θ for which a sufficient statistic exists. If, in the reduced forms (6-14) and (6-18), we write $x - \theta$ for x, we shall obtain the p.d.f.'s of the location-parameter families of distributions admitting a single sufficient statistic for the location parameter θ. We now proceed to discuss the location-parameter families with reduced forms (6-14) and (6-18).

6.3 Location-parameter Family of Distributions with Reduced Form $f_1(x) = \exp[c_1 \exp(c_2x) + c_3x + c_4]$ [Equation (6-14)] when $c_2 \neq 0$

Since the range of the distribution (6-14) is $(-\infty, \infty)$, the constants c_1, c_2, c_3, and c_4 are subject to the restriction

$$\int_{-\infty}^{\infty} f_1(x)\, dx = \int_{-\infty}^{\infty} \exp[c_1 \exp(c_2x) + c_3x + c_4]\, dx = 1 \qquad (c_2 \neq 0). \qquad (6\text{-}19)$$

It can be shown that the infinite integral in Eq. (6-19) converges if, and only if, $c_1 < 0$ and $c_2c_3 > 0$, in which case Eq. (6-19) gives

$$\exp(c_4) = \frac{|c_2|\, (-c_1)^{c_3/c_2}}{\Gamma(c_3/c_2)} \qquad (6\text{-}20)$$

It may be noted that the transformation $y = -c_1 \exp(c_2x)$ transforms Eq. (6-14) into the Gamma distribution with parameter c_3/c_2. Since c_4 is determined in terms of c_1, c_2, and c_3 from Eq. (6-20), the number of independent arbitrary constants in Eq. (6-14) is three. Writing $x - \theta$ for x in Eq. (6-14), the p.d.f. of the location-parameter family of distributions with reduced form (6-14) is

$$f(x;\, \theta) = f_1(x - \theta) = \exp\{c_1 \exp[c_2(x - \theta)] + c_3(x - \theta) + c_4\}$$
$$(-\infty < x < \infty,\ -\infty < \theta < \infty,\ c_1 < 0,\ c_2c_3 > 0). \qquad (6\text{-}21)$$

Since Eq. (6-21) can be written in the form (2-6) (the c's of the other parameters being supposed known), the statistic $T = \sum_{i=1}^{n} \exp(c_2x_i)$ is sufficient for the location parameter θ, where $x_1, \ldots, x_n$ is a random sample of n members from a distribution of the family (6-21). The family of distributions (6-21) belongs to the "iterated exponential" type, which is also known by other names: "double-exponential distributions" or "extreme-value distributions." Since the number of essential parameters in Eq. (6-14) is three (viz., c_1, c_2 and c_3), it may appear at first that the number of essential parameters in Eq. (6-21) [which is obtained by writing $x - \theta$ for x in Eq. (6-14)] is four (viz., θ, c_1, c_2, and c_3). This is, however, not the case; the number of essential parameters in Eq. (6-21) is still three, i.e., writing $x - \theta$ for x in the reduced form (6-14) does not alter the number of essential parameters. To show this, consider the term $c_1 \exp c_2(x - \theta)$ in Eq. (6-21). Since $c_1 < 0$ and $-\infty < \theta < \infty$, we note that

$$c_1 \exp c_2(x - \theta) = (-c_1)\ [-\exp(c_2 x)]\ \exp(-c_2\theta). \qquad (6\text{-}22)$$

Now $-c_1$ is a positive arbitrary constant, and $\exp(-c_2\theta)$ is also a positive arbitrary constant. Hence we can drop $-c_1$ and retain $\exp(-c_2\theta)$ without any loss of generality in Eq. (6-22). That is, we can replace $c_1 \exp[c_2(x - \theta)]$ by $-\exp[c_2(x - \theta)]$. Hence Eq. (6-21) becomes

$$f(x;\ \theta) = \exp\{-\exp[c_2(x - \theta)] + c_3(x - \theta) + c_4\}$$
$$(-\infty < x < \infty,\ -\infty < \theta < \infty,\ c_2 c_3 > 0), \qquad (6\text{-}23)$$

which shows that Eq. (6-23) depends on three essential parameters only (viz., θ, c_2, and c_3).

Writing $k_1 = c_2$, $k_2 = c_3$, and $k_3 = c_4$, Eq. (6-23) may be written as

$$f(x;\ \theta) = \exp\{-\exp[k_1(x - \theta)] + k_2(x - \theta) + k_3\}$$
$$(-\infty < x < \infty,\ -\infty < \theta < \infty,\ k_1 k_2 > 0). \qquad (6\text{-}24)$$

Further it is possible to rewrite Eq. (6-24) into two elegant forms, each form exhibiting the presence of a scale parameter σ (supposed to be known in this case) and a numerical parameter λ. For we consider the following two subcases of Eq. (6-24).

Subcase 1: $k_1 > 0$, so that $k_2 > 0$ also.

Write $k_1 = 1/\sigma$ $(\sigma > 0)$ and $k_2 = \lambda/\sigma$ $(\lambda > 0)$. Then Eq. (6-24) becomes

$$f(x;\ \theta) = \exp\left[-\exp\left(\frac{x - \theta}{\sigma}\right) + \lambda\left(\frac{x - \theta}{\sigma}\right) + k_3\right]$$

$$(-\infty < x < \infty,\ -\infty < \theta < \infty,\ \sigma > 0,\ \lambda > 0). \qquad (6\text{-}25)$$

In Eq. (6-25) $\int_{-\infty}^{\infty} f(x;\ \theta)\ dx = 1$ gives

$$\exp(k_3) = \frac{1}{\sigma\Gamma(\lambda)}. \qquad (6\text{-}26)$$

Hence $$f(x;\ \theta) = \frac{1}{\sigma\Gamma(\lambda)} \exp\left[-\exp\frac{x - \theta}{\sigma} + \lambda\left(\frac{x - \theta}{\sigma}\right)\right]$$

$$(-\infty < x < \infty,\ -\infty < \theta < \infty,\ \sigma > 0,\ \lambda > 0). \qquad (6\text{-}27)$$

It may be noted that in Eq. (6-27) $y = \exp[(x - \theta)/\sigma]$ has the Gamma distribution with parameter λ, and $z = \theta + \sigma\ \exp[(x - \theta)/\sigma]$ has the nonregular Pearson Type III distribution (range: θ to ∞) with θ as a location parameter, σ as a scale parameter, and λ as a numerical parameter.

Subcase 2: $k_1 < 0$, so that $k_2 < 0$ also.

Write $k_1 = -1/\sigma$ $(\sigma > 0)$ and $k_2 = -\lambda/\sigma$ $(\lambda > 0)$. We have $\exp(k_3) = 1/[\sigma\ \Gamma(\lambda)]$. Then Eq. (6-24) becomes

$$f(x;\ \theta) = \frac{1}{\sigma\Gamma(\lambda)} \exp\left[-\exp - \frac{x - \theta}{\lambda} - \lambda\left(\frac{x - \theta}{\sigma}\right)\right]$$

$$(-\infty < x < \infty,\ -\infty < \theta < \infty,\ \sigma > 0,\ \lambda > 0). \qquad (6\text{-}28)$$

In Eq. (6-28) $y = \exp[-(x - \theta)/\lambda]$ has the Gamma distribution with parameter λ, and $z = \theta + \sigma\ \exp[(x - \theta)/\sigma]$ has the nonregular Pearson Type III distribution (range: $-\infty$ to θ) with θ, σ, and λ being respectively the location, the scale, and the numerical parameter.

In Eq. (6-28), if we put $\theta = 0$, $\sigma = 1$, and $\lambda = 1$, we get what Gumbel (1958) calls the "first double-exponential distribution,"

$$f(x) = \exp(-e^{-x} - x) \quad (-\infty < x < \infty). \tag{6-29}$$

In Eq. (6-27), if we put $\theta = 0$, $\sigma = 1$, and $\lambda = 1$, we get Gumbel's "second double-exponential distribution,"

$$f(x) = \exp(-e^{x} + x) \quad (-\infty < x < \infty). \tag{6-30}$$

We may call Eqs. (6-28) and (6-27) the generalized first and second iterated exponential distributions.

6.4 Location-parameter Family of Distributions with Reduced Form $f_1(x) = \exp(c_1x^2 + c_2x + c_3)$ [Equation (6-18)]

Since the range of the distribution (6-18) is $(-\infty, \infty)$, the constants c_1, c_2, and c_3 are subject to the restriction

$$\int_{-\infty}^{\infty} f_1(x)\,dx = \int_{-\infty}^{\infty} \exp(c_1x^2 + c_2x + c_3)\,dx = 1. \tag{6-31}$$

The infinite integral in Eq. (6-31) converges if, and only if, $c_1 < 0$, in which case Eq. (6-31) gives

$$\exp(c_3) = \left(-\frac{c_1}{\pi}\right)^{1/2} \exp\left(\frac{c_2^2}{4c_1}\right). \tag{6-32}$$

The distribution (6-18) is normal with mean $-c_2/2c_1$ and variance $(-1/2c_1)^{1/2}$. Writing $x - \theta$ for x in Eq. (6-18) the p.d.f. of the location-parameter family of distributions with reduced form (6-18) is

$$f(x;\theta) = f_1(x-\theta) = \exp[c_1(x-\theta)^2 + c_2(x-\theta) + c_3]$$
$$(-\infty < x < \infty,\ -\infty < \theta < \infty,\ c_1 < 0,\ -\infty < c_2 < \infty). \tag{6-33}$$

For the distribution (6-33), the sample mean $\bar{x}$ is sufficient for θ (the c's of the other parameters being supposed known). The number of essential parameters in Eq. (6-33) is two, which is the same as the number of essential parameters in Eq. (6-18). To prove this, we may write

$$\frac{1}{2\sigma^2} = -c_1 \qquad (\sigma > 0),$$

and replace $\theta - c_2/2c_1$ by θ, without any loss of generality. Then Eq. (6-18) assumes the well-known standard form of a normal distribution,

$$f(x;\ \theta) = \frac{1}{\sigma\sqrt{2\pi}} \exp\left[-\frac{(x-\theta)^2}{2\sigma^2}\right]$$

$$(-\infty < x < \infty,\ -\infty < \theta < \infty,\ \sigma > 0), \qquad (6\text{-}34)$$

where the scale parameter σ is supposed known while considering a single sufficient statistic for θ.

In Eq. (6-34) θ is now the mean of the family of normal distributions, σ is the standard deviation, and the distributions are symmetrical about θ.

6.5 Conclusion

We have shown that the family of normal distributions (with known scale parameter) and the family of generalized iterated exponential distributions (with known scale and numerical parameters) are the only location-parameter families of distributions admitting a single sufficient statistic for the location parameter, under the assumptions stated in Sec. 6.1. The main assumption of Sec. 6.1 is the existence of the third derivative of the p.d.f. of the location-parameter family of distributions.

The two families, the iterated exponential and the normal, may be further differentiated from each other by their properties that the normal distribution is symmetrical about the location parameter if the mean is taken as the location parameter, whereas the iterated exponential distribution is skewed whatever may be the choice of the location parameter.

7. Scale-Parameter Families of Distributions of the Regular Type Admitting a Single Sufficient Statistic for the Scale Parameter

7.1 Statement of the Problem and the Assumptions

Let $f(x; \sigma) = (1/\sigma)\ f_1(x/\sigma)$ be the p.d.f. of a scale-parameter family of distributions of the regular type. As shown in Sec. 4.2, the range of admissible values of σ is the open infinite interval $\sigma > 0$; and as shown in Sec. 4.4, the range R of the family of distributions is either $(-\infty, \infty)$ or $(-\infty, 0)$ or $(0, \infty)$. We wish to determine $f(x; \sigma)$ so that a single sufficient statistic exists for σ for all $\sigma > 0$ and for all random samples of any size $n \geq 2$.

Assumptions. We assume the Koopman regularity conditions for the general form (2-6) for existence of a single sufficient statistic for a single unknown parameter σ in the regular case. The Koopman regularity conditions imply mainly the existence of the second derivative $f_1''(x)$. We further assume the existence of the third derivative $f_1'''(x)$.

7.2 Solution of the Problem in Section 7.1

Under the assumptions of Sec. 7.1, we have [using Eq. (2-6)]

$$f(x; \sigma) = \frac{1}{\sigma} f_1\left(\frac{x}{\sigma}\right) = \exp[u(\sigma)\ v(x) + A(x) + B(\sigma)] \qquad (7\text{-}1)$$

for all $x \in R$ and for all $\sigma > 0$. Writing $1/\sigma$ for σ in Eq. (7-1), we have

$$\sigma\ f_1(\sigma x) = \exp\left[u\left(\frac{1}{\sigma}\right)\ v(x) + A(x) + B\left(\frac{1}{\sigma}\right)\right],$$

i.e.,

$$f_1(\sigma x) = \exp[u_1(\sigma)\ v(x) + A(x) + B_1(\sigma)], \qquad (7\text{-}2)$$

where $u_1(\sigma) = u(1/\sigma)$ and $B_1(\sigma) = B(1/\sigma) - \log \sigma$.

$$\log f_1(\sigma x) = u_1(\sigma)\ v(x) + A(x) + B_1(\sigma). \qquad (7\text{-}3)$$

Write $\quad f_2(x) = \log f_1(x). \qquad (7\text{-}4)$

Then Eq. (7-3) becomes

$$f_2(\sigma x) = u_1(\sigma)\ v(x) + A(x) + B_1(\sigma). \qquad (7\text{-}5)$$

Differentiating Eq. (7-5) with respect to x,

$$\sigma\ f_2'(\sigma x) = u_1(\sigma)\ v'(x) + A'(x),$$

i.e.,

$$\sigma\ x\ f_2'(\sigma x) = u_1(\sigma)\ x\ v'(x) + x\ A'(x),$$

i.e.,

$$\sigma\ x\ f_2'(\sigma x) = u_1(\sigma)\ v_1(x) + A_1(x), \qquad (7\text{-}6)$$

where $v_1(x)$ and $A_1(x)$ are functions of x.

Write $f_3(x) = x\ f_2'(x)$. (7-7)

Then Eq. (7-6) becomes

$$f_3(\sigma x) = u_1(\sigma)\ v_1(x) + A_1(x). \qquad (7\text{-}8)$$

Differentiating Eq. (7-8) with respect to σ,

$$x\ f_3'(\sigma x) = u_1'(\sigma)\ v_1(x),$$

i.e., $f_3'(\sigma x) = u_1'(\sigma)\ \dfrac{v_1(x)}{x}$,

i.e., $f_3'(\sigma x) = u_2(\sigma)\ v_2(x)$. (7-9)

Write $f_4(x) = f_3'(x)$, (7-10)

so that Eq. (7-9) becomes

$$f_4(\sigma x) = u_2(\sigma)\ v_2(x). \qquad (7\text{-}11)$$

Equation (7-11) is a functional equation satisfied by $f_4(x)$ for all $x \in R$ and for all $\sigma > 0$. $f_4'(x)$ exists, since $f''_1{}'(x)$ exists. Hence, using the results of Sec. 2.6.2, the general solution of Eq. (7-11) is

$$f_4(x) = c_1\ x^{c_2}, \qquad (7\text{-}12)$$

where c_1 and c_2 are arbitrary constants. From Eqs. (7-10) and (7-12),

$$f_3'(x) = c_1 x^{c_2}. \tag{7-13}$$

Two cases arise:

CASE I: $c_2 \neq -1$.

In this case the solution of Eq. (7-13) is

$$f_3(x) = \frac{c_1}{c_2 + 1} x^{c_2+1} + c_3, \tag{7-14}$$

where c_3 is an arbitrary constant. From Eqs. (7-7) and (7-14),

$$x\, f_2'(x) = \frac{c_1}{c_2 + 1} x^{c_2+1} + c_3$$

i.e.,
$$f_2'(x) = \frac{c_1}{c_2 + 1} x^{c_2} + \frac{c_3}{x}. \tag{7-15}$$

Solving Eq. (7-15),

$$f_2(x) = \frac{c_1}{(c_2 + 1)^2} x^{c_2+1} + c_3 \log|x| + c_4 \quad (c_2 \neq -1), \tag{7-16}$$

where c_4 is an arbitrary constant. Replacing the set of arbitrary constants c_1, c_2, c_3, and c_4 by the set of arbitrary constants k_1, k_2 $(k_2 \neq 0)$, k_3, and k_4, we have

$$f_2(x) = k_1 x^{k_2} + k_3 \log|x| + k_4 \quad (k_2 \neq 0). \tag{7-17}$$

Hence, from Eqs. (7-4) and (7-17),

$$f_1(x) = \exp(k_1 x^{k_2} + k_3 \log|x| + k_4) \quad (k_2 \neq 0). \tag{7-18}$$

CASE II: $c_2 = -1$ in Equation (7-13), $f_3'(x) = c_1 x^{c_2}$.

In this case Eq. (7-13) becomes

$$f_3'(x) = \frac{c_1}{x},$$

which gives

$$f_3(x) = c_1 \log|x| + c_2, \qquad (7\text{-}19)$$

where c_2 is an arbitrary constant. Equations (7-7) and (7-19) give

$$x\, f_2'(x) = c_1 \log|x| + c_2,$$

i.e.,
$$f_2'(x) = c_1 \frac{\log|x|}{x} + \frac{c_2}{x}. \qquad (7\text{-}20)$$

Solving Eq. (7-20),

$$f_2(x) = \frac{1}{2} c_1 (\log|x|)^2 + c_2 \log|x| + c_3,$$

where c_3 is an arbitrary constant. Replacing $\frac{1}{2}c_1$ by c_1, we have

$$f_2(x) = c_1 (\log|x|)^2 + c_2 \log|x| + c_3. \qquad (7\text{-}21)$$

From Eqs. (7-4) and (7-21),

$$f_1(x) = \exp[c_1 (\log|x|)^2 + c_2 \log|x| + c_3]. \qquad (7\text{-}22)$$

We have thus shown that if $f_1(x)$ satisfies the functional equation (7-1), then $f_1(x)$ must satisfy either Eq. (7-18) or Eq. (7-22). Conversely, it is easily verified that if $f_1(x)$ satisfies either Eq. (7-18) or Eq. (7-22), it satisfies Eq. (7-1) also. Hence Eqs. (7-18) and (7-22) constitute the general solution of the functional equation (7-1). Since $f_1(x)$ is the p.d.f. of the reduced form of the family of distributions with p.d.f. $f(x; \sigma) = (1/\sigma) f_1(x/\sigma)$, we have thus shown that Eqs. (7-18) and (7-22) give the p.d.f. of the reduced forms of the scale-parameter families of distributions admitting a single sufficient statistic for the scale parameter. We now proceed to discuss briefly the scale-parameter families with reduced forms (7-18) and (7-22).

7.3 Scale-parameter Family of Distributions with Reduced Form $f_1(x) = \exp(k_1 x^{k_2} + k_3 \log|x| + k_4)$ [Equation (7-18)] when $k_2 \neq 0$

Since the range of the distribution (7-18) is R, where R is as defined in Sec. 7.1, the constants k_1, k_2, k_3, and k_4 are subject to the restriction $\int_R f_1(x)\,dx = 1$. For example, if $R = (-\infty, \infty)$, then $\int_{-\infty}^{\infty} f_1(x)\,dx$ converges if $k_1 < 0$ and k_2 is a positive even integer, and then k_4 is determined from $\int_{-\infty}^{\infty} f_1(x)\,dx = 1$. But if k_2 is an odd integer, $\int_{-\infty}^{\infty} f_1(x)\,dx$ is not convergent. It is interesting to note that Eq. (7-18) gives also the normal distribution when $k_1 < 0$, $k_2 = 2$, and $k_3 = 0$.

If $R = (0, \infty)$, Eq. (7-18) gives a Gamma distribution when $k_1 < 0$, $k_2 = 1$, and $k_3 > -1$.

The scale-parameter family of distributions with reduced form (7-18) is

$$f(x;\sigma) = \frac{1}{\sigma} f_1\left(\frac{x}{\sigma}\right) = \frac{1}{\sigma} \exp\left[k_1\left(\frac{x}{\sigma}\right)^{k_2} + k_3 \log\left|\frac{x}{\sigma}\right| + k_4\right]$$

$$(x \in R,\ \sigma > 0,\ k_2 \neq 0). \quad (7\text{-}23)$$

Since Eq. (7-23) can be written in the form (2-6) (the c's of the other parameters being supposed known), the statistic $T = \sum_{i=1}^{n} x_i^k$ is sufficient for σ.

The number of essential parameters in Eq. (7-23) is three. The term $k_1(x/\sigma)^{k_2}$ can be replaced by $-(x/\sigma)^{k_2}$ without any loss of generality, since $k_1 < 0$ and $\sigma > 0$. The k's can be renumbered and Eq. (7-23) can be rewritten in the form

$$f(x;\sigma) = \frac{k_3}{\sigma} \exp\left[-\left(\frac{x}{\sigma}\right)^{k_1} + k_2 \frac{\log|x|}{\sigma}\right]$$

$$(x \in R,\ \sigma > 0,\ k_1 \neq 0), \quad (7\text{-}24)$$

where σ, k_1, and k_2 are the essential parameters in Eq. (7-24), and k_3 is determined from the condition $\int_R f(x;\sigma)\,dx = 1$.

7.4 Scale-parameter Family of Distributions with Reduced Form $f_1(x) = \exp[c_1(\log|x|)^2 + c_2 \log|x| + c_3]$ [Equation (7-22)]

Since the range of the distribution (7-22) is R, the distribution is symmetrical about the origin if $R = (-\infty, \infty)$. $\int_{-\infty}^{\infty} f_1(x)\, dx$ converges if $c_1 < 0$. If $R = (0, \infty)$, it is interesting to note that $Y = \log X$ has a normal distribution.

The scale-parameter family of distributions with reduced form (7-22) is

$$f(x;\sigma) = \frac{1}{\sigma} f_1\left(\frac{x}{\sigma}\right) = \frac{1}{\sigma} \exp\left[c_1 \frac{(\log|x|)^2}{\sigma^2} + c_2 \frac{\log|x|}{\sigma} + c_3\right]$$

$$(x \in R,\ \sigma > 0,\ c_1 < 0). \qquad (7\text{-}25)$$

Here the statistic $T = \prod_{i=1}^{n} x_i$ is sufficient for σ when the other parameters c's are known. It may be noted, however, that the number of essential parameters in Eq. (7-25) is two (including σ), and Eq. (7-25) can be rewritten accordingly.

7.5 Conclusion

We have shown that Eqs. (7-24) and (7-25) give the only scale-parameter families of distributions admitting a single sufficient statistic for the scale parameter, under the assumptions stated in Sec. 7.1. The main assumption of Sec. 7.1 is the existence of the third derivative of the p.d.f. of the scale-parameter family of distributions. The family (7-24) includes the normal distribution (with known location) as a particular case.

It may be noted that the families (7-24) and (7-25) can be derived from the corresponding location-parameter families (6-21) and (6-33) by writing log x for x, and log σ for θ, in virtue of the relationship between location and scale parameters pointed out in Sec. 4.8. However, as remarked in Sec. 4.8, this relationship unnecessarily restricts the range of the scale-parameter family of distributions to the interval $(0, \infty)$ in the regular case. The results obtained in the present section are free from this restriction.

8. The Normal Family of Distributions As the Only Location-Scale-Parameter Family of Distributions of the Regular Type Admitting A Minimal Pair of Jointly Sufficient Statistics for the Parameters of Location and Scale

8.1 Statement of the Problem and the Assumptions

Let $f(x; \theta, \sigma) = (1/\sigma)\ f_1[(x - \theta)/\sigma]$ be the p.d.f. of a location-scale-parameter family of distributions of the regular type. As shown in Sec. 5.3, R_1 is the range of the family of distributions. We wish to determine $f(x; \theta, \sigma)$ so that a minimal pair of jointly sufficient statistics exists for (θ, σ) for all $\theta \in R_1$, $\sigma > 0$, and all random samples of any size $n \geq 3$.

Assumptions. We assume the Koopman regularity conditions for the general form (2-7) for existence of a minimal pair of jointly sufficient statistics for a pair of unknown parameters. The Koopman regularity conditions imply mainly the existence of the second-order partial derivatives $\partial^2 f/(\partial\theta\ \partial x)$ and $\partial^2 f/(\partial\sigma\ \partial x)$, which is equivalent to the existence of $f_1''(x)$. We further assume the existence of the third derivative $f_1'''(x)$.

8.2 The Normal Family of Distributions as the Unique Solution of the Problem in Section 8.1

Under the assumptions of Sec. 8.1, we have [using Eq. (2-7)]

$$f(x; \theta, \sigma) = \frac{1}{\sigma}\ f_1\left(\frac{x - \theta}{\sigma}\right) = \exp[u_1(\theta, \sigma)\ v_1(x) + u_2(\theta, \sigma)\ v_2(x) + A(x) + B(\theta, \sigma)] \tag{8-1}$$

for all $x \in R_1$, $\theta \in R_1$, and $\sigma > 0$.

Writing $-\theta$ for θ, and $1/\sigma$ for σ in Eq. (8-1),

$$\sigma\ f_1(\sigma x + \sigma\theta) = \exp\left[u_1\left(-\theta, \frac{1}{\sigma}\right) v_1(x) + u_2\left(-\theta, \frac{1}{\sigma}\right) v_2(x) + A(x) + B\left(-\theta, \frac{1}{\sigma}\right)\right]$$

i.e.,
$$f_1(\sigma x + \sigma\theta) = \exp[u_{11}(\theta, \sigma)\ v_1(x) + u_{21}(\theta, \sigma)\ v_2(x) + A(x) + B_1(\theta, \sigma)], \tag{8-2}$$

where u_{11}, u_{21}, and B_1 are functions of (θ, σ). We have

$$\log f_1(\sigma x + \sigma\theta) = u_{11}(\theta, \sigma)\ v_1(x) + u_{21}(\theta, \sigma)\ v_2(x) + A(x) + B_1(\theta, \sigma). \tag{8-3}$$

Write $f_2(x) = \log f_1(x)$, (8-4)

so that Eq. (8-3) becomes

$$f_2(\sigma x + \sigma\theta) = u_{11}(\theta, \sigma)\ v_1(x) + u_{21}(\theta, \sigma)\ v_2(x) + A(x) + B_1(\theta, \sigma). \tag{8-5}$$

Differentiating Eq. (8-5) partially with respect to x,

$$\sigma\ f_2'(\sigma x + \sigma\theta) = u_{11}(\theta, \sigma)\ v_1'(x) + u_{21}(\theta, \sigma)\ v_2'(x) + A'(x). \tag{8-6}$$

Differentiating Eq. (8-6) partially with respect to θ,

$$\sigma^2\ f_2''(\sigma x + \sigma\theta) = \frac{\partial\ u_{11}(\theta, \sigma)}{\partial\theta}\ v_1'(x) + \frac{\partial\ u_{21}(\theta, \sigma)}{\partial\theta}\ v_2'(x),$$

i.e., $$f_2''(\sigma x + \sigma\theta) = u_{12}(\theta, \sigma)\ v_{11}(x) + u_{22}(\theta, \sigma)\ v_{21}(x), \tag{8-7}$$

where u_{12} and u_{22} are functions of (θ, σ), and v_{11} and v_{21} are functions of x.

Write $f_3(x) = f_2''(x)$. (8-8)

From Eqs. (8-4) and (8-8),

$$f_3(x) = \frac{d^2}{dx^2}[\log f_1(x)]. \tag{8-9}$$

Using Eq. (8-8), Eq. (8-7) becomes

$$f_3(\sigma x + \sigma\theta) = u_{12}(\theta, \sigma)\ v_{11}(x) + u_{22}(\theta, \sigma)\ v_{21}(x). \tag{8-10}$$

Equation (8-10) is a functional equation satisfied by $f_3(x)$ for all $x \in R_1$, $\theta \in R$, and $\sigma > 0$. Since $f_1'''(x)$ exists, it follows from Eq. (8-9) that $f_3'(x)$ exists. The functional equation (8-10) is the same as the functional equation (2-24). Hence, using Eq. (2-25), the general solution of Eq. (8-10) is

$$f_3(x) = \lambda_1 x + \lambda_2, \tag{8-11}$$

where λ_1 and λ_2 are arbitrary constants. From Eqs. (8-9) and (8-11),

$$\frac{d^2}{dx^2} \log[f_1(x)] = \lambda_1 x + \lambda_2, \tag{8-12}$$

which gives $$f_1(x) = \exp(c_0 x^3 + c_1 x^2 + c_2 x + c_3), \tag{8-13}$$

where c_0, c_1, c_2, and c_3 are arbitrary constants.

We have thus shown that if $f_1(x)$ satisfies the functional equation (8-1), then $f_1(x)$ must satisfy Eq. (8-13). Conversely, it is easily verified that if $f_1(x)$ satisfies Eq. (8-13), it satisfies Eq. (8-1) also. Hence Eq. (8-13) gives the general solution of the functional equation (8-1).

Now the range of the distribution (8-13) is $R_1 = (-\infty, \infty)$, and the constants c_0, c_1, c_2, and c_3 are subject to the restriction $\int_{-\infty}^{\infty} f_1(x)\, dx = 1$. Since $\int_{-\infty}^{\infty} f_1(x)\, dx$ converges if, and only if, $c_0 = 0$ and $c_1 < 0$, we have

$$f_1(x) = \exp(c_1 x^2 + c_2 x + c_3) \qquad (-\infty < x < \infty,\ c_1 < 0). \tag{8-14}$$

Equation (8-14) is the p.d.f. of the reduced form of the location-scale-parameter family of distributions admitting a minimal pair of jointly sufficient statistics for the pair (θ, σ) of location and scale parameters. Since Eq. (8-14) is a normal distribution, it follows that the normal family of distributions is the only location-scale-parameter family of distributions of the regular type admitting a minimal pair of jointly sufficient statistics for the parameters of location and scale for all admissible values of the location and scale parameters and for all random samples of any size $n \geq 3$ under the assumptions stated in Sec. 8.1. The main assumption of Sec. 8.1 is the existence of the third derivative of the p.d.f. of the location-scale-parameter family of distributions.

9. Location-Parameter Families of Distributions of the Nonregular Type Admitting a Single Sufficient Statistic for the Location Parameter

9.1 Statement of the Problem and the Assumptions

Let $f(x;\ \theta) = f_1(x - \theta)$ be the p.d.f. of a location-parameter family of distributions of the nonregular type with ranges $[a(\theta), b(\theta)]$. As shown in Sec. 3.2, the range of admissible values of θ is R_1. As shown in Sec. 3.4, the ranges of the family of distributions belong to one of the three types: Eqs. (3-6) - (3-8).

$$R = \text{the union of the ranges } [a(\theta), b(\theta)] \text{ for all } \theta \in R_1$$
$$= R_1.$$

We wish to determine $f(x;\ \theta)$ so that a single sufficient statistic exists for θ for all $\theta \in R_1$ and for all random samples of any size $n \geq 2$.

It follows from Eqs. (3-6) - (3-8) that $a(\theta)$ and $b(\theta)$ are continuous, differentiable, monotonic functions of θ, and $a(\theta) < b(\theta)$, for all $\theta \in R_1$. Thus the assumptions mentioned in Case I of Sec. 2.5 are satisfied.

Assumption. The only assumption we make is the existence of the derivative $(\partial/\partial\theta)\ f(x;\ \theta)$, which is equivalent to the assumption of the existence of the first derivative $f_1'(x)$.

9.2 Solution of the Problem in Section 9.1

Since the assumptions of Case I of Sec. 2.5 are satisfied, it follows that the necessary and sufficient conditions for the existence of a single sufficient statistic for θ for all $\theta \in R_1$ and for all random samples of any size $n \geq 2$ are

1. $a(\theta)$ and $b(\theta)$ are monotonic in θ in opposite senses, and (9-1)

2. $f(x;\ \theta) = f_1(x - \theta) = g(x)\ h(\theta)$ for all $x \in R_1$ and $\theta \in R_1$. (9-2)

In virtue of Eqs. (3-6) - (3-8), the condition (9-1) is satisfied if, and only if, the ranges are either $(-\infty, b_0 + \theta)$ or $(a_0 + \theta, \infty)$.

To solve the functional equation (9-2), put $-\theta$ for θ, so that

$$f_1(x + \theta) = g(x)\ h(-\theta),$$

i.e., $$f_1(x+\theta) = g(x)\ h_1(\theta). \tag{9-3}$$

Equation (9-3) is a functional equation satisfied by $f_1(x)$ for all $x \varepsilon R_1$ and $\theta \varepsilon R_1$. Also $f_1'(x)$ exists by the assumption in Sec. 9.1. The functional equation (9-3) is the same as the functional equation (2-12). Hence, using Eq. (2-13), the general solution of Eq. (9-3) is

$$f_1(x) = c_1 \exp(c_2 x), \tag{9-4}$$

where c_1 and c_2 are arbitrary constants.

We have thus shown that Eq. (9-4) is the p.d.f. of the reduced form of the location-parameter family of distributions admitting a single sufficient statistic for the location parameter, and that the ranges of the family of distributions are either $(-\infty, b_0 + \theta)$ or $(a_0 + \theta, \infty)$, where a_0 and b_0 are constants. We now proceed to discuss the location-parameter families with reduced form (9-4) and ranges $(-\infty, b_0 + \theta)$ or $(a_0 + \theta, \infty)$.

9.3 Location-parameter Family of Distributions with Reduced Form $f_1(x) = c_1 \exp(c_2 x)$ [Equation (9-4)] and Ranges $(-\infty, b_0 + \theta)$

In this case the range of the reduced distribution (9-4) is $(-\infty, b_0)$. The constants b_0, c_1, and c_2 are subject to the restriction

$$\int_{-\infty}^{b_0} f_1(x)\ dx = \int_{-\infty}^{b_0} c_1 \exp(c_2 x)\ dx = 1. \tag{9-5}$$

The infinite integral in Eq. (9-5) converges if, and only if, $c_2 > 0$, in which case we have

$$c_1 = c_2 \exp(-c_2 b_0). \tag{9-6}$$

Hence

$$f_1(x) = c_2 \exp[c_2(x - b_0)] \quad (-\infty < x \le b_0,\ c_2 > 0), \tag{9-7}$$

or, replacing c_2 by $c > 0$, we have

$$f_1(x) = c \exp[-c(b_0 - x)]$$
$$(-\infty < x \le b_0 + \theta,\ c > 0,\ -\infty < \theta < \infty). \tag{9-8}$$

It may be noted that the number of essential parameters in Eq. (9-8) is two, viz., θ and c. Without any loss of generality, we may replace $b_0 + \theta$ by θ itself, i.e., we may put $b_0 = 0$. Hence Eq. (9-8) becomes

$$f(x;\ \theta) = c \exp[-c(\theta - x)]$$
$$(-\infty < x \le \theta,\ -\infty < \theta < \infty,\ c > 0). \tag{9-9}$$

The known positive parameter c in Eq. (9-9) may be replaced by $1/\sigma$, where σ is a known scale parameter in Eq. (9-9). Hence Eq. (9-9) becomes

$$f(x;\ \theta) = \frac{1}{\sigma} \exp\left(-\frac{\theta - x}{\sigma}\right)$$
$$(-\infty < x \le \theta,\ -\infty < \theta < \infty,\ \sigma > 0). \tag{9-10}$$

Equation (9-10) is the exponential family of distributions with ranges $(-\infty, \theta)$, where θ is an unknown location parameter and σ is a known scale parameter. In this case the largest member in the sample is a sufficient statistic for θ.

9.4 Location-parameter Family of Distributions with Reduced Form $f_1(x) = c_1 \exp(c_2 x)$ Equation (9-4) and Ranges $(a_0 + \theta, \infty)$

The discussion for this case is exactly on similar lines as the discussion for the case in Sec. 9.3. In this case the location-parameter family of distributions can be written in the form

$$f(x;\ \theta) = \frac{1}{\sigma} \exp\left(-\frac{x - \theta}{\sigma}\right)$$
$$(\theta \le x < \infty;\ -\infty < \theta < \infty,\ \sigma > 0). \tag{9-11}$$

Equation (9-11) is the exponential family of distributions with ranges (θ, ∞), where θ is an unknown location parameter and σ is a known scale parameter. In this case the smallest member in the sample is a sufficient statistic for θ.

9.5 Conclusion

We have shown, under the assumption of existence of the first derivative of the p.d.f. of a location-parameter family of distributions, that the two exponential families of distributions with ranges $(-\infty, \theta)$ and (θ, ∞) (where θ is the location parameter, and the scale parameter is supposed known) are the only location-parameter families of distributions of the nonregular type admitting a single sufficient statistic for the location parameter θ for all $\theta \in R_1$ and for all random samples of any size $n \geq 2$.

It may be noted that the nonregular case discussed in Sec. 9 belongs to the pure type (Sec. 2.3).

10. Scale-Parameter Families of Distributions of the Non-regular Type Admitting a Single Sufficient Statistic for the Scale Parameter

10.1 Statement of the Problem and the Assumptions

Let $f(x; \sigma) = (1/\sigma)\, f_1(x/\sigma)$ be the p.d.f. of a scale-parameter family of distributions of the nonregular type with ranges $[a(\sigma), b(\sigma)]$. As shown in Sec. 4.2, the range of admissible values of σ is the open interval $(0, \infty)$. As shown in Sec. 4.4, the ranges of the family of distributions belong to one of the five types: Eqs. (4-5) - (4-9).

R is the union of the ranges $[a(\sigma), b(\sigma)]$ for all $\sigma > 0$. Note that

$$
\begin{aligned}
R = (-\infty, \infty) &= R_1 && \text{if } a_0 < 0,\ b_0 > 0, \\
&= (0, \infty) && \text{if } a_0 \geq 0, \\
&= (-\infty, 0) && \text{if } b_0 \leq 0.
\end{aligned}
$$

We wish to determine $f(x; \sigma)$ so that a single sufficient statistic exists for σ for all $\sigma > 0$ and for all random samples of any size $n \geq 2$.

It follows from Eqs. (4-5) - (4-9) that $a(\sigma)$ and $b(\sigma)$ are continuous, differentiable, monotonic functions of σ, and $a(\sigma) < b(\sigma)$ for all $\sigma > 0$. Thus the assumptions mentioned in Case I of Sec. 2.5 are satisfied.

Assumption. The only assumption we make is the existence of the derivative $(\partial/\partial\sigma)\, f(x; \sigma)$, which is equivalent to the assumption of the existence of the first derivative $f_1'(x)$.

10.2 Solution of the Problem in Section 10.1

Since the assumptions of Case I of Sec. 2.5 are satisfied, it follows that the necessary and sufficient conditions for the existence of a single sufficient statistic for σ for all $\sigma > 0$ and for all random samples of any size $n \geq 2$ are

1. $a(\sigma)$ and $b(\sigma)$ are monotonic in σ in opposite senses, and (10-1)

2. $f(x; \sigma) = \frac{1}{\sigma} f_1\left(\frac{x}{\sigma}\right) = g(x)\, h(\sigma)$ for all $x \in R$ and $\sigma > 0$. (10-2)

In virtue of Eqs. (4-5) - (4-9), the condition (10-1) is satisfied if, and only if, the ranges of the family of distributions belong to one of the five types:

Type 1: $(a_0\sigma, b_0\sigma)$ $(a_0 < 0, b_0 > 0)$, (10-3)

Type 2: $(-\infty, b_0\sigma)$, (10-4)

Type 3: $(0, b_0\sigma)$ $(b_0 > 0)$, (10-5)

Type 4: $(a_0\sigma, \infty)$, (10-6)

Type 5: $(a_0\sigma, 0)$ $(a_0 < 0)$, (10-7)

where a_0 and b_0 are finite. In Eqs. (10-4) and (10-6) a_0 and b_0 must be nonzero.

To solve the functional equation (10-2), put $1/\sigma$ for σ, so that

$$\sigma f_1(\sigma x) = g(x)\, h\left(\frac{1}{\sigma}\right),$$

i.e., $$f_1(\sigma x) = g(x)\, \frac{1}{\sigma}\, h\left(\frac{1}{\sigma}\right),$$

i.e., $$f_1(\sigma x) = g(x)\, h_1(\sigma). \tag{10-8}$$

Equation (10-8) is a functional equation satisfied by $f_1(x)$ for all $x \in R$ and $\sigma > 0$. Also $f_1'(x)$ exists by the assumption of Sec. 10.1. The functional equation (10-8) is the same as the functional equation (2-14). Hence, using Eq. (2-15), the general solution of Eq. (10-8) is

$$f_1(x) = c_1 x^{c_2}, \tag{10-9}$$

where c_1 and c_2 are arbitrary constants.

We have thus shown that Eq. (10-9) is the p.d.f. of the reduced form of the scale-parameter family of distributions admitting a single sufficient statistic for the scale parameter, and that the ranges of the family of distributions belong to one of the five types: Eqs. (10-3) - (10-7).

We now proceed to discuss briefly the scale-parameter family of distributions with reduced form (10-9) and ranges belonging to one of the five types (10-3) - (10-7).

10.3 Scale-parameter Family of Distributions with Reduced Form $f_1(x) = c_1 x^{c_2}$ [Equation (10-9)] and Ranges $(a_0\sigma, b_0\sigma)$, where $a_0 < 0$, $b_0 > 0$, and a_0 and b_0 are Finite

In this case the range of the reduced distribution (10-9) is (a_0, b_0). The constants a_0, b_0, c_1, and c_2 are subject to the restrictions that $f_1(x) \geq 0$ for all x in (a_0, b_0), and

$$\int_{a_0}^{b_0} f_1(x)\, dx = \int_{a_0}^{b_0} c_1 x^{c_2}\, dx = 1. \tag{10-10}$$

X assumes both positive and negative values. $f_1(x)$ becomes infinite at $x = 0$ if $c_2 < 0$. In particular, if c_2 is a nonnegative even integer, the condition $f_1(x) \geq 0$ over (a_0, b_0) is satisfied, and the condition (10-10) gives

$$c_1 = \frac{c_2+1}{b_0^{c_2+1} - a_0^{c_2+1}} \qquad (10\text{-}11)$$

The p.d.f. of the family of scale-parameter distributions with reduced form (10-9) and ranges $(a_0\sigma, b_0\sigma)$ $(a_0 < 0, b_0 > 0)$ is

$$f(x;\ \sigma) = \frac{1}{\sigma} f_1\left(\frac{x}{\sigma}\right) = \frac{c_1}{\sigma} \left(\frac{x}{\sigma}\right)^{c_2}$$

$$(a_0\sigma \leq x \leq b_0\sigma,\ a_0 < 0,\ b_0 > 0,\ \sigma > 0), \qquad (10\text{-}12)$$

where c_1 is given by Eq. (10-11). In this case, when c_2 is known, the single statistic

$$T = \max\left[\frac{x_{(1)}}{a_0}, \frac{x_{(n)}}{b_0}\right], \qquad (10\text{-}13)$$

is sufficient for σ, where $x_{(1)}$ and $x_{(n)}$ are respectively the smallest and the greatest members in the sample. An important special case is provided by $c_2 = 0$, in which case Eq. (10-12) gives a rectangular distribution with p.d.f.

$$f(x;\ \sigma) = \frac{1}{(b_0 - a_0)\sigma}$$

$$(a_0\sigma \leq x \leq b_0\sigma,\ a_0 < 0,\ b_0 > 0,\ \sigma > 0). \qquad (10\text{-}14)$$

Here $(b_0 - a_0)\sigma = R$ is the range of the rectangular distribution.

10.4 Scale-parameter Family of Distributions with Reduced Form $f_1(x) = c_1 x^{c_2}$ [Equation (10-9)] and Ranges $(-\infty, b_0\sigma)$, Where b_0 is Finite and Nonzero

In this case the range of the reduced distribution (10-9) is $(-\infty, b_0)$.

$$\int_{-\infty}^{b_0} f_1(x)\, dx = \int_{-\infty}^{b_0} c_1 x^{c_2}\, dx$$

converges if, and only if, $b_0 < 0$ and $c_2 < -1$. We consider those values of $c_2 < -1$ which make $x^{c_2} > 0$. For example, we may take c_2 to be a negative even integer ≤ -2, or we may take $c_2 = p/q$, where p is an even integer.

$$\text{Then} \qquad c_1 = \frac{c_2+1}{b_0^{c_2+1}}, \tag{10-15}$$

$$\text{and} \qquad f_1(x) = \frac{(c_2+1)x^{c_2}}{b_0^{c_2+1}\sigma}$$

$$(-\infty < x \leq b_0,\ b_0 < 0,\ c_2 < -1), \tag{10-16}$$

$$f(x;\ \sigma) = \frac{1}{\sigma} f_1\left(\frac{x}{\sigma}\right) = \frac{c_2+1}{b_0^{c_2+1}} \left(\frac{x}{\sigma}\right)^{c_2}$$

$$(-\infty < x \leq b_0,\ b_0 < 0,\ c_2 < -1,\ \sigma > 0). \tag{10-17}$$

Without any loss of generality we may put $b_0 = -1$, and noting that $(-1)^{c_2} = +1$, and $c_2+1 = -c$ $(c > 0)$, we have

$$f(x;\ \sigma) = \frac{c}{\sigma} \left(\frac{\sigma}{x}\right)^{c+1} \qquad (-\infty < x \leq -\sigma,\ c > 0,\ \sigma > 0). \tag{10-18}$$

Here the largest member in the sample is a sufficient statistic for σ when c is known.

10.5 Scale-parameter Family of Distributions with Reduced Form $f_1(x) = c_1 x^{c_2}$ [Equation (10-9)] and Ranges $(0, b_0\sigma)$, Where b_0 is finite and positive

In this case the range of the reduced distribution (10-9) is $(0, b_0)$. Here $f_1(x)$ becomes infinite at $x = 0$ if $c_2 < 0$, but

$\int_0^{b_0} f_1(x)\,dx$ converges at $x = 0$ if $c_2 > -1$. We may therefore allow all real values > -1, and take the nonnegative value of x^{c_2} if c_2 is not an integer. We have

$$c_1 = \frac{c_2 + 1}{b_0^{c_2 + 1}}, \tag{10-19}$$

and
$$f_1(x) = \frac{c_2 + 1}{b_0^{c_2 + 1}} x^{c_2}$$

$$(0 \leq x \leq b_0,\ b_0 > 0,\ c_2 > -1), \tag{10-20}$$

$$f(x;\ \sigma) = \frac{1}{\sigma} f_1\left(\frac{x}{\sigma}\right) = \frac{c_2 + 1}{b_0^{c_2 + 1}\sigma} \left(\frac{x}{\sigma}\right)^{c_2}$$

$$(0 < x < b_0\sigma,\ b_0 > 0,\ c_2 > -1,\ \sigma > 0). \tag{10-21}$$

Without any loss of generality we may put $b_0 = 1$, and replace c_2 by $c > -1$. Hence

$$f(x;\ \sigma) = \frac{c + 1}{\sigma} \left(\frac{x}{\sigma}\right)^{c} \qquad (0 \leq x \leq \sigma,\ \sigma > 0,\ c > -1). \tag{10-22}$$

Here the largest member in the sample is sufficient for σ when c is known.

If in Eq. (10-22) we put $c = 0$, we get the rectangular distribution

$$f(x;\ \sigma) = \frac{1}{\sigma} \qquad (0 \leq x \leq \sigma,\ \sigma > 0). \tag{10-23}$$

10.6 Scale-parameter Family of Distributions with Reduced Form $f_1(x) = c_1x^{c_2}$ [Equation (10-9)] and Ranges $(a_0\sigma, \infty)$, Where a_0 is Finite and Nonzero

In this case, the range of the reduced distribution (10-9) is (a_0, ∞).

$$\int_{a_0}^{\infty} f_1(x)\, dx = \int_{a_0}^{\infty} c_1 x^{c_2}\, dx$$

converges if, and only if, $a_0 > 0$ and $c_2 < -1$. Then

$$c_1 = -\frac{c_2 + 1}{a_0^{c_2 + 1}}, \tag{10-24}$$

$$f_1(x) = -\frac{c_2 + 1}{a_0^{c_2 + 1}} x^{c_2}$$

$$(a_0 \le x < \infty,\ a_0 > 0,\ c_2 < -1), \tag{10-25}$$

$$f(x;\ \sigma) = \frac{1}{\sigma} f_1\left(\frac{x}{\sigma}\right) = -\frac{c_2 + 1}{a_0^{c_2 + 1}\sigma} \left(\frac{x}{\sigma}\right)^{c_2}$$

$$(a_0\sigma \le x < \infty,\ a_0 > 0,\ c_2 < -1,\ \sigma > 0). \tag{10-26}$$

Without any loss of generality, we may put $a_0 = 1$, $c_2 + 1 = -c$ $(c > 0)$. Hence

$$f(x;\ \sigma) = \frac{c}{\sigma} \left(\frac{\sigma}{x}\right)^{c+1} \quad (\sigma \le x < \infty,\ \sigma > 0,\ c > 0). \tag{10-27}$$

Here the smallest member in the sample is a sufficient statistic for σ when c is known.

10.7 Scale-parameter Family of Distributions with Reduced Form $f_1(x) = c_1 x^{c_2}$ [Equation (2-9)] and Ranges $(a_0\sigma, 0)$, Where a_0 is Negative and Finite

In this case the range of the reduced distribution is $(a_0, 0)$. Here $x = 0$ is the only possible singular point of $f_1(x)$ and $\int_{a_0}^{0} f_1(x)\, dx$ converges at $x = 0$ if, and only if, $c_2 > -1$. We may therefore consider those values of $c_2 > -1$ for which x^{c_2} is nonnegative. For example, x^{c_2} is nonnegative if c_2 is a nonnegative even integer, or if $c_2 = p/q$, where p is even.

Then
$$c_1 = -\frac{c_2 + 1}{a_0^{c_2 + 1}}, \tag{10-28}$$

and
$$f_1(x) = -\frac{c_2 + 1}{a_0^{c_2 + 1}} x^{c_2}$$
$$(a_0 \le x \le 0,\ a_0 < 0,\ c_2 > -1), \tag{10-29}$$

$$f(x;\ \sigma) = \frac{1}{\sigma} f_1\left(\frac{x}{\sigma}\right) = -\frac{c_2 + 1}{a_0^{c_2 + 1}\sigma}\left(\frac{x}{\sigma}\right)^{c_2}$$
$$(a_0\sigma \le x \le 0,\ a_0 < 0,\ \sigma > 0,\ c_2 > -1). \tag{10-30}$$

Without any loss of generality, we may put $a_0 = -1$ and $c = c_2 > -1$. Hence

$$f(x;\ \sigma) = \frac{c + 1}{\sigma}\left(\frac{x}{\sigma}\right)^{c} \quad (-\sigma \le x \le 0,\ \sigma > 0,\ c_2 > -1). \tag{10-31}$$

Here the smallest member in the sample is a sufficient statistic for σ when c is known. If in Eq. (10-31) we put $c = 0$, we get the rectangular distribution

$$f(x;\ \sigma) = \frac{1}{\sigma} \quad (-\sigma \le x \le 0,\ \sigma > 0). \tag{10-32}$$

10.8 Conclusion

We have shown, under the assumption of existence of the first derivative of the p.d.f. of a scale-parameter family of distributions, that the only scale-parameter families of distributions of the nonregular type admitting a single sufficient statistic for the scale parameter σ for all $\sigma > 0$ and for all random samples of any size $n \ge 2$ are those given by Eqs. (10-12), (10-18), (10-22), (10-27), and (10-31). The families (10-12), (10-22), and (10-31) include also the scale-parameter families of rectangular distributions as special cases.

It may be noted that the nonregular case of Sec. 10 belongs to the pure type (Sec. 2.3).

11. The Rectangular and the Exponential Families of Distributions As the Only Location Scale-Parameter Families of Distributions of the Nonregular Type Admitting a Minimal Pair of Jointly Sufficient Statistics for the Parameters of Location and Scale

11.1 Statement of the Problem and the Assumptions

Let $f(x;\ \theta,\ \sigma) = 1/\sigma\ f_1[(x - \theta)/\sigma]$ be the p.d.f. of a location-scale-parameter family of distributions of the nonregular type with ranges $[a(\theta, \sigma), b(\theta, \sigma)]$. As shown in Sec. 5.3, the ranges of the family of distributions belong to one of the seven types: Eqs. (5-6) - (5-12).

R is the union of the ranges $[a(\theta, \sigma), b(\theta, \sigma)]$ for all $\theta \in R_1$ and all $\sigma > 0$. We note that $R = R_1$.

We wish to determine $f(x;\ \theta,\ \sigma)$ so that a minimal pair of jointly sufficient statistics exists for (θ, σ) for all $\theta \in R_1$ and $\sigma > 0$ for all random samples of any size $n \geq 3$.

As shown in Eqs. (5-6) - (5-12) the ranges $[a(\theta, \sigma), b(\theta, \sigma)]$ depend on either both parameters (θ, σ) or on the single parameter θ. Accordingly we make two cases for the solution of the problem in question, and we make different assumptions in the two cases.

CASE I: The ranges depend on both parameters of location and scale.

Assumption. The only assumption we make is the existence of the first derivative $f_1'(x)$. This assumption implies that $(\partial/\partial x)\ f(x;\ \theta,\ \sigma)$, and $(\partial/\partial\theta)\ f(x;\ \theta,\ \sigma)$ and $(\partial/\partial\sigma)\ f(x;\ \theta,\ \sigma)$ exist.

CASE II: The ranges depend on the single parameter of location only.

As shown in Sec. 5.3, the ranges of the family of distributions in this case are either $(-\infty, \theta)$ or (θ, ∞). It follows that $a(\theta)$ and $b(\theta)$ are continuous, differentiable, and monotonic in θ, and $a(\theta) < b(\theta)$ for all $\theta \in R_1$. These assumptions mentioned in Subcase 1 or Case II of Sec. 2.5 are therefore satisfied.

Assumption. We assume the Koopman regularity conditions for the general form (2-8) for the regular case with $p = 2$ parameters and $\nu = 1$ statistic. It may be noted that the Koopman regularity conditions involve the assumption of existence of $\partial^2 f/(\partial\theta\ \partial x)$ and $\partial^2 f/(\partial\sigma\ \partial x)$, which is equivalent to the assumption of existence of $f_1''(x)$.

11.2 The Rectangular Family of Distributions as the Unique Solution of the Problem in Section 11.1 when the Ranges are of the Case I Type

When the ranges depend on both parameters (θ, σ), it follows from Subcase 1 of Case II of Sec. 2.5 that a necessary and sufficient condition for the existence of a minimal pair of jointly sufficient statistics for (θ, σ) for all $\theta \varepsilon R_1$, for all $\sigma > 0$, and for all random samples of any size $n \geq 3$ is that

$$f(x; \theta, \sigma) = \frac{1}{\sigma} f_1\left(\frac{x - \theta}{\sigma}\right) = g(x)\ h(\theta, \sigma) \qquad (11\text{-}1)$$

for all $x \varepsilon R_1$ and for all $\theta \varepsilon R_1$ and $\sigma > 0$.

Writing $-\theta$ for θ, and $1/\sigma$ for σ in Eq. (11-1)

$$\sigma\ f_1(\sigma x + \sigma\theta) = g(x)\ h\left(-\theta, \frac{1}{\sigma}\right),$$

i.e., $$f_1(\sigma x + \sigma\theta) = g(x)\ \frac{1}{\sigma}\, h\left(-\theta, \frac{1}{\sigma}\right),$$

i.e., $$f_1(\sigma x + \sigma\theta) = g(x)\ h_1(\theta, \sigma). \qquad (11\text{-}2)$$

Equation (11-2) is a functional equation satisfied by $f_1(x)$ for all $x \varepsilon R_1$, $\theta \varepsilon R_1$, and $\sigma > 0$. Also $f_1'(x)$ exists by assumption. The functional equation (11-2) is the same as the functional equation (2-32). Hence, using Eq. (2-33), the general solution of Eq. (11-2) is

$$f_1(x) = c, \qquad (11\text{-}3)$$

where c is an arbitrary constant.

We have shown that if $f_1(x)$ satisfies the functional equation (11-1), then $f_1(x)$ must satisfy Eq. (11-3). Conversely, it is easily verified that if $f_1(x)$ satisfies Eq. (11-3), it satisfies Eq. (11-1) also. Hence Eq. (11-3) gives the general solution of the functional equation (11-1).

Let (a_0, b_0) be the range of the reduced distribution with p.d.f. (11-3). It follows that a_0 and b_0 must both be finite, since

$$\int_{a_0}^{b_0} f_1(x)\ dx = \int_{a_0}^{b_0} c\ dx = c(b_0 - a_0) = 1. \qquad (11\text{-}4)$$

We have $c = \dfrac{1}{b_0 - a_0}$, (11-5)

and $f_1(x) = \dfrac{1}{b_0 - a_0} \quad (a_0 \le x \le b_0)$. (11-6)

Equation (11-6) gives the p.d.f. and the range of the reduced form of the only location-scale-parameter family of distributions of the nonregular type with ranges depending on both location and scale parameters (θ, σ), and admitting a minimal pair of jointly sufficient statistics for (θ, σ). Since Eq. (11-6) is a rectangular distribution, it follows that the rectangular family of distributions is the only location-scale-parameter family of distributions of the nonregular type with ranges depending on both location and scale parameters (θ, σ) and admitting a minimal pair of jointly sufficient statistics for (θ, σ) for all $\theta \in R_1$, for all $\sigma > 0$, and for all random samples of any size $n \geq 3$. The rectangular family of distributions with reduced form (11-6) is then given by

$$f(x;\ \theta,\ \sigma) = \frac{1}{\sigma} f_1\left(\frac{x - \theta}{\sigma}\right) = \frac{1}{(b_0 - a_0)\sigma}$$

$$(\theta + a_0\sigma \le x \le \theta + b_0\sigma). \qquad (11\text{-}7)$$

Here the pair $[x_{(1)}, x_{(n)}]$ of the smallest and the greatest members in the sample is minimal sufficient for (θ, σ).

The two numerical constants a_0 and b_0 in Eq. (11-7) are at our disposal. One way of fixing them up is to take the center μ and the range R as the pair of location and scale parameters, so that

$$\mu = \theta + \frac{1}{2}(a_0 + b_0)\sigma, \qquad (11\text{-}8)$$

and $R = (b_0 - a_0)\sigma$. (11-9)

Equations (11-8) and (11-9) give the terminals of the ranges:

$$\theta + a_0\sigma = \mu - \frac{1}{2} R, \qquad (11\text{-}10)$$

and $\theta + b_0\sigma = \mu + \frac{1}{2} R$. (11-11)

Hence Eq. (11-7) becomes

$$f(x;\ \mu,\ R) = \frac{1}{R} \quad \left(\mu - \frac{1}{2} R \leq x \leq \mu + \frac{1}{2} R\right). \tag{11-12}$$

11.3 The Exponential Family of Distributions as the Unique Solution of the Problem in Section 11.1 when the Ranges are of the Case II Type

In this case the ranges of the family of distributions are either $(-\infty,\ \theta)$ or $(\theta,\ \infty)$. It follows from Subcase 2 of Case II of Sec. 2.5 that necessary and sufficient conditions for the existence of a minimal pair of jointly sufficient statistics for $(\theta,\ \sigma)$ for all $\theta \in R_1$, for all $\sigma > 0$, and for all random samples of any size $n \geq 3$ are

1. $a(\theta)$ and $b(\theta)$ are monotonic in θ in opposite senses, and (11-13)

2. $f(x;\ \theta,\ \sigma) = \frac{1}{\sigma} f_1\left(\frac{x - \theta}{\sigma}\right) = \exp[u(\theta,\ \sigma)\ v(x) + A(x) + B(\theta,\ \sigma)]$ for all $x \in R_1$, $\theta \in R_1$, and $\sigma > 0$. (11-14)

Since the ranges $(-\infty,\ \theta)$ and $(\theta,\ \infty)$ satisfy the condition (11-13), it follows that the necessary and sufficient condition is Eq. (11-14) only. To solve the functional equation (11-14), put $-\theta$ for θ, and $1/\sigma$ for σ, so that Eq. (11-14) becomes

$$\sigma\ f_1(\sigma x + \sigma\theta) = \exp\left[u\left(-\theta,\ \frac{1}{\sigma}\right)\ v(x) + A(x) + B\left(-\theta,\ \frac{1}{\sigma}\right)\right],$$

i.e.,
$$f_1(\sigma x + \sigma\theta) = \exp[u_1(\theta,\ \sigma)\ v(x) + A(x) + B_1(\theta,\ \sigma)], \tag{11-15}$$

where u_1 and B_1 are functions of $(\theta,\ \sigma)$. We have

$$\log f_1(\sigma x + \sigma\theta) = u_1(\theta,\ \sigma)\ v(x) + A(x) + B_1(\theta,\ \sigma). \tag{11-16}$$

Write
$$f_2(x) = \log f_1(x), \tag{11-17}$$

so that
$$f_2(\sigma x + \sigma\theta) = u_1(\theta,\ \sigma)\ v(x) + A(x) + B_1(\theta,\ \sigma). \tag{11-18}$$

Differentiating Eq. (11-18) partially with respect to x,

$$\sigma f_2'(\sigma x + \sigma\theta) = u_1(\theta, \sigma) v'(x) + A'(x),$$

i.e., $$f_2'(\sigma x + \sigma\theta) = \frac{1}{\sigma} u_1(\theta, \sigma) v'(x) + \frac{1}{\sigma} A'(x),$$

i.e., $$f_2'(\sigma x + \sigma\theta) = u_{11}(\theta, \sigma) v_1(x) + u_{12}(\theta, \sigma) v_2(x), \quad (11\text{-}19)$$

where u_{11} and u_{12} are functions of (θ, σ), and $v_1(x)$ and $v_2(x)$ are functions of x. Write

$$f_3(x) = f_2'(x). \quad (11\text{-}20)$$

From Eqs. (11-17) and (11-20),

$$f_3(x) = \frac{d}{dx} [\log f_1(x)]. \quad (11\text{-}21)$$

Using Eq. (11-20), Eq. (11-19) becomes

$$f_3(\sigma x + \sigma\theta) = u_{11}(\theta, \sigma) v_1(x) + u_{12}(\theta, \sigma) v_2(x). \quad (11\text{-}22)$$

Equation (11-22) is a functional equation satisfied by $f_3(x)$ for all $x \in R_1$, $\theta \in R_1$, and $\sigma > 0$. Also, since $f_1''(x)$ exists, it follows from Eq. (11-21) that $f_3'(x)$ exists. The functional equation (11-22) is exactly the same as the functional equation (2-24), hence, using Eq. (2-25), the general solution of Eq. (11-22) is

$$f_3(x) = \lambda_1 x + \lambda_2, \quad (11\text{-}23)$$

where λ_1 and λ_2 are arbitrary constants. Equations (11-21) and (11-23) give

$$\frac{d}{dx} [\log f_1(x)] = \lambda_1 x + \lambda_2,$$

i.e., $$f_1(x) = \exp c_0 x^2 + c_1 x + c_2 , \quad (11\text{-}24)$$

where c_0, c_1, and c_2 are arbitrary constants. We have shown that if $f_1(x)$ satisfies Eq. (11-14), then $f_1(x)$ satisfies Eq. (11-24). Conversely, it is easily verified that if $f_1(x)$ satisfies Eq. (11-24), then $f_1(x)$ satisfies Eq. (11-14) if, and only if, $c_0 = 0$. Hence the general solution of Eq. (11-14) is

$$f_1(x) = \exp(c_1 x + c_2),$$

i.e., $f_1(x) = c \exp(c_1 x)$, (11-25)

where c and c_1 are arbitrary constants.

Since Eq. (11-25) is the reduced form of the p.d.f. of an exponential family of distributions, it follows that the exponential family of distributions is the only location-scale-parameter family of distributions of the nonregular type with ranges depending on only one (viz., the location) of the two parameters and admitting a minimal pair of jointly sufficient statistics for the pair (θ, σ) of location and scale parameters for all $\theta \in R_1$, for all $\sigma > 0$, and for all random samples of any size $n \geq 3$. We derive two subfamilies of the family of exponential distributions corresponding to the ranges $(-\infty, \theta)$ and (θ, ∞):

11.3.1 Exponential subfamily with ranges $(-\infty, \theta)$ and reduced form (11-25), $f_1(x) = c \exp(c_1 x)$

In this case the range of the reduced distribution (11-25) is $(-\infty, 0)$. The constants c and c_1 are subject to the restrictions that $f_1(x) \geq 0$, and

$$\int_{-\infty}^{0} f_1(x)\, dx = c \int_{-\infty}^{0} \exp(c_1 x)\, dx = 1.$$

We must have $c > 0$ and $c = c_1$ for the convergence of $\int_{-\infty}^{0} \exp(c_1 x)\, dx$. Hence Eq. (11-25) becomes

$$f_1(x) = c\, e^{cx} \qquad (-\infty < x \leq 0). \tag{11-26}$$

The exponential subfamily of distributions with ranges $(-\infty, \theta)$ and reduced form (11-26) is then given by

$$f(x; \theta, \sigma) = \frac{1}{\sigma} f_1\left(\frac{x - \theta}{\sigma}\right) = \frac{c}{\sigma} \exp\left[c\left(\frac{x - \theta}{\sigma}\right)\right]$$
$$(-\infty < x \leq \theta,\ -\infty < \theta < \infty,\ \sigma > 0,\ c > 0). \tag{11-27}$$

Since $\sigma > 0$ and $c > 0$, the constant c is no longer essential in Eq. (11-27). Without any loss of generality, we may replace c/σ by σ, i.e., we may put $c = 1$. Then Eq. (11-27) becomes

$$f(x;\ \theta,\ \sigma) = \frac{1}{\sigma} \exp\left(-\frac{\theta - x}{\sigma}\right)$$

$$(-\infty < x \le \theta,\ -\infty < \theta < \infty,\ \sigma > 0). \qquad (11\text{-}28)$$

In this case the largest member in the sample and the sample mean are a minimal pair of jointly sufficient statistics for (θ, σ).

11.3.2 Exponential subfamily with ranges (θ, ∞) and reduced form (11-25), $f_1(x) = c \exp(c_1 x)$

In this case the range of the reduced distribution (11-25) is $(0, \infty)$, and

$$\int_0^\infty f_1(x)\ dx = \int_{00}^\infty c \exp(c_1 x)\ dx$$

converges to 1 if, and only if, $-c_1 = c > 0$. Then Eq. (11-25) becomes

$$f_1(x) = c\, e^{-cx} \qquad (0 \le x < \infty). \qquad (11\text{-}29)$$

The exponential subfamily of distributions with ranges (θ, ∞) and reduced form (11-29) is

$$f(x;\ \theta,\ \sigma) = \frac{c}{\sigma} \exp\left[-\frac{c}{\sigma}(x - \theta)\right]$$

$$(\theta \le x < \infty,\ -\infty < \theta < \infty,\ \sigma > 0,\ c > 0). \qquad (11\text{-}30)$$

Without any loss of generality, we may put $c = 1$ in Eq. (11-30). Hence

$$f(x;\ \theta,\ \sigma) = \frac{1}{\sigma} \exp\left(-\frac{x - \theta}{\sigma}\right)$$

$$(\theta \le x < \infty,\ -\infty < \theta < \infty,\ \sigma > 0). \qquad (11\text{-}31)$$

In this case the smallest member in the sample and the sample mean are a minimal pair of jointly sufficient statistics for (θ, σ).

11.4 Concluding Remarks

It should be noted that the nonregular Case I of Sec. 11.1 belongs to the pure type, whereas the nonregular Case II of Sec. 11.1 belongs to the mixed type (Sec. 2.3). In the pure-type nonregular case, the solution is obtained under the assumption of the existence of only the first derivative of the p.d.f. of the location-scale-parameter family of distributions. In the mixed-type nonregular case the solution is obtained under the assumption of the existence of the second derivative of the family of distributions.

ACKNOWLEDGMENTS

I wish to express my sincerest thanks to the authorities of the Iowa State University and to Professor T. A. Bancroft, Director of the Statistical Laboratory, for help and facilities for research work.

This work was supported by the National Science Foundation (N.S.F. GP-1155) and was written when the author was Fulbright Visiting Professor at Iowa State University.

REFERENCES

Darmois, G. (1935): Sur les lois de probabilité à estimation exhaustive, C. R. Acad. Sci. Paris 200, 1265.

Gumbel, E. J. (1958): Statistics of Extremes, New York, Columbia Univ. Press.

Huzurbazar, V. S. (1964): The general forms of distributions admitting sufficient statistics for parameters in nonregular cases, J. R. Stat. Soc. B

Koopman, B. O. (1936): On distributions admitting a sufficient statistic, Trans. Amer. Math. Soc. 39, 399.

Levi-Civita, T. (1913): Sulle funzioni che ammettono una formula d'addizione del tipo $f(x + y) = \sum_{1}^{n} X_i(x)\ Y_i(y)$, Rendiconti 22, 2° sem., 181.

Pexider, Von, H. W. (1902): Verallgemeinerung gewisser Cauchy'schen functional-gleichungen, Monatschefte Math. Phys. 14, 293.

Pitman, E. J. G. (1936): Sufficient statistics and intrinsic accuracy, Proc. Camb. Phil. Soc. 32, 567.

Stackel, P. (1913): Sulla equazione funzionale $f(x + y) = \sum_{i=1}^{n} X_i(x)\ Y_i(y)$, Rendiconti 22, 2° sem. 392.

APPENDIX I

Chronological list of relevant papers and books published after the monographs were written:

1. Jeffreys, H. (1961): Theory of Probability, 3rd ed., London, Oxford Univ. Press.

2. Huzurbazar, V. S. (1966): Some invariants of some discrete distributions admitting sufficient statistics for parameters, Proc. Intern. Symp. Classical and Contagious Discrete Distributions, Montreal (Aug. 1963), p. 231 (Pergamon Press). Reprinted in Sankhya A, 28, 215 (1966).

3. Huzurbazar, V. S. (1967): Invariants of probability distributions, Presidential Address, Statistics Section, 54th Session of Indian Science Congress Association, Hyderabad (Jan. 1967).

4. Huzurbazar, V. S. (1968): Inverse probability and confidence intervals, Contributions in Statistics and Agricultural Sciences, Presented to Dr. V. G. Panse on His 62nd Birthday, Indian Society of Agricultural Statistics, p. 93.

5. Huzurbazar, V. S. (1971): Invariant Forms for the prior probability of parameters of the bivariate normal distribution, Poona Univ. J. Sci. Tech., no. 40, 219.

6. Sudakov, V. N. (1971): A note on marginal sufficiency, Sov. Math. Dokl. 12, no. 3, 796.

7. Huzurbazar, V. S. (1972): Explicit sampling distribution and the posterior probability distribution of the maximum likelihood estimators of parameters of a regular-type distribution admitting sufficient statistics for parameters, Poona Univ. J. Sci. Tech., no. 41, 109.

8. Huzurbazar, V. S. (1973): Inverse probability and the use of transformations, Poona Univ. J. Sci. and Tech., no. 44, 153.

9. Rao, C. R. (1973): Linear Statistical Inference and Its Applications, 2nd ed., New York, Wiley.

APPENDIX II

Sufficiency and Huzurbazar's Conjecture.

Let $X_1, \ldots, X_n$ be i.i.d. r.v. with distribution function F_θ, $\theta \in \Theta$. According to definition, a measurable function $T = f(X_1, \ldots, X_n)$ is a sufficient statistic if the conditional distribution of any other statistic Y given T is independent of θ. Huzurbazar conjectured that to establish sufficiency of T it is enough to examine only marginal sufficiency, i.e., the conditional distribution of X_i given T is in independent of θ for any i. A partial proof of this proposition is given by J. K. Ghosh, and a more complete proof of a somewhat more general statement is given by V. N. Sudakov ["A note on marginal sufficiency", Doklady, p. 198], 1971.

AUTHOR INDEX

SUBJECT INDEX